JOHN D. ROBSON
COLIN J. DODDS
DONALD B. MACVEAN
VINCENT R. PALING
UNIVERSITY OF GLASGOW

RANDOM VIBRATIONS

COURSE HELD AT THE DEPARMENT
OF GENERAL MECHANICS
OCTOBER 1971

UDINE 1971

SPRINGER - VERLAG   WIEN – NEW YORK

ISBN 3-211-81223-7-Springer-Verlag Wien-New York
ISBN 0-387-81223-7 Springer-Verlag New York-Wien

PREFACE

*The subject of Random Vibration is of great signi-
ficance for engineers and other workers in the field
of mechanics, for sources of irregularly fluctuating
forces exist in many practical situations. Yet the formal training of such potential users hardly permits of
its inclusion and the acquisition of the necessary know
ledge, both for newly qualified and experienced workers,
must depend on private study or postgraduate teaching.*

*It has seemed to us in Glasgow that in this subject
private study is best supplemented by postgraduate teach
ing, and we have presented an introductory course of a-
bout twenty lectures designed to give a sound basis for
understanding. Naturally such a course can only be introductory and can hope only to give a basis on which
further study can build. But the concepts of random
processes are of some subtlety, and the analytical meth
ods of description and response are of some complica-
tion. At first some guided introduction to the subject
is almost essential, and even a course of twenty lectures concentrated into one week can be invaluable here.*

*The success of such a course at Glasgow provided
the motivation for presentation of this course at Udine. The CISM timetable does not easily lend itself to
the inclusion of one-week concentrated courses, but
this difficulty was overcome. The course as presented*

here is self-contained as an introduction, but was arranged to precede more advanced lectures in Random Vibration problems given by other workers, in order to make these more comprehensible to continuing students. It seems to have served well in both purposes.

The opportunity to give the lectures a more permanent form is welcome. Though they were designed primarily for verbal delivery the collected notes do provide a continuous narrative which may be read as a whole by those with no previous knowledge of the subject. There is introductory material on statistical analysis and even on vibration theory: yet there is some account of particular applications, and of computation and analysis.

The authors are grateful to Professor Olszak and Professor Sobrero for their interest, and for offering the opportunity to present the course at Udine, and also to many others at CISM whose cooperation made the course possible, not least to Signora Bertozzi.

Udine, September 1970

J.D. Robson

Chapter 1   INTRODUCTION

1.1. The nature of the problem

The subject of random vibration arises from the need to relate the response of a structure to an excitation which – by its nature – is not amenable to precise description. The designer of an aircraft, for example, must take account of the stresses which arise from runway roughness or air turbulence, although the imposed displacements or pressures vary with time in a manner which defies analytical description, and which indeed cannot even be known until the aircraft goes into service. Such problems arise in many forms in all branches of engineering, and orthodox vibration analysis gives no help in their solution. New techniques and concepts are required if such random ly varying excitations are to be treated.

If nothing at all is known about an excitation it will of course be quite impossible to predict response. Nor can _complete_ prediction of response be expected unless a similarly complete description of excitation is available. But in many practical problems something less than a complete description of response is quite sufficient, and for this an incomplete specification of excitation may prove quite adequate.

Consider for example a number of identical aircraft which fly at the same speed through uniformly turbulent air. The pressures experienced by each aircraft would not be identical: pressures at corresponding points could each be plotted as a function of time, and very different records would emerge from each measuring point. And yet one could reasonably expect these different turbulence histories to be generally similar in their essential characteristics, and to have very similar effects in their capacity to damage the aircraft or their contents. There must in such a case be some similarity in the pressures experienced, and equally in the stresses arising from them, and thus in the damage resulting. If we can isolate whatever there is in common between such records, and can describe it quantitatively, we can then proceed to problems of response determination, and so build up a theory of Random Vibration.

1.2. The problem of description

It will be helpful in considering the essential similarity of a number of records to indicate in what ways records can be dissimilar. Consider

(a) the four records (a), (b), (c), (d) shown, each of which may be taken

(b) to represent the variation of pressure with time. Records (a) and (b)

(c) are quite different in detail: they are not identical. Yet they do have

something in common, particularly (d)
when they are both contrasted with
record (c) where the fluctuations
are much more rapid, and with record (d) where the amplitudes
are much greater.

The way to description lies – not surprisingly –
through statistics, by means of which the idiosyncrasies of individual records can be disregarded and what they have in general can be described. If, for example, we were to devise a
crude statistical classification based on mean values of amplitude and of times between zero crossings, we should find that
(a) and (b) were in the same category but separated from (c)
and (d). Something much more sophisticated than this is required, but this is the way in which we shall have to think.

What we have to devise is a description precise
enough to be useful, yet imprecise enough to be applicable to
each randomly-varying quantity generated by a single physical
process. Fortunately such a description can be found. It does
moreover prove amenable to response analysis.

In the above it has been tacitly assumed that all
records, and all the processes generating them, do possess a certain consistency; that a quantity which at first varies in the
manner of (a) or (b) will not gradually change so as to take
the form of (c) or (d), or that a given process will not sometimes give rise to records of the form of (a) and sometimes of

(c) or (d). Such good behaviour is not a natural characteristic of all randomly varying quantities in practice, and in assuming it to occur we are limiting the scope of our analysis. Such consistency, however – or stationarity as we shall later call it – is often enough approached in practice to form a useful basis for analysis, in much the same way as the assumption of a linear stress-strain law proves useful in stress analysis.

1.3. The nature of the subject

When we are able to describe excitations, making use not only of statistical ideas but also of harmonic analysis, it will be found that response determination is not in principle a difficult matter. Analysis can be developed which, once excitations are described and system properties are known, enables the response to be determined, or at least described in the same manner and to the same extent as the excitation. The establishment of such response relationships will clearly be an important part of our subject.

But analysis of response is not enough. It has been convenient to introduce the subject of random vibration in terms of response determination, but many problems arise which are not in themselves connected to the response problem. The response of a system, whether expressed in terms of displacement, acceleration, or stress, is rarely sufficient to indicate whether the system will perform satisfactorily in service. We shall need to look into the _effects_ of random vibration on materials or systems, and this will lead us to look more deeply into the implications of our statistical descriptions. Problems arise in instrumentation, or in testing, which are not simple response determinations, although they necessarily require an understanding of the basic techniques of description and analysis.

And we shall eventually need to consider the extension of our techniques to more sophisticated conditions where, for example, stationarity and linearity may not be permissible assumptions.

1.4. The nature of the course

The nature of the course follows from the nature of the subject. We must establish a suitable technique of _description_ and understand what it implies. We must develop techniques of _response analysis_ and consider the implications of the results. We must take some typical practical _applications_, and show how our techniques can be usefully applied to practical problems. The essential ideas of statistics and generalized harmonic analysis will be introduced as the need arises in the lectures on description: the essential ideas of orthodox _vibration theory_ will be presented as they become necessary in a group of three lectures.

There are therefore four main groups of lectures, but it must be noted that the groups are not arranged consecutively. The time table has been arranged so that each lecture is presented where it fits in most conveniently; and problems of random vibration are introduced as early as possible in the course.

It will be convenient to restrict our attention strictly to vibration problems, though the analysis of the response of systems to random excitation finds its application also in other fields, and many results will have a wider relevance.

Bibliography for Chapter 1

The following are the principal books covering the subject of Random Vibration:-

1.1   Crandall S.H. and Mark W.D.: Random Vibration in Mechanic al Systems: Academic Press 1963.
Elementary account of the subject, restricted in scope, but with extended discussion of behaviour of two-freedom system, and some attention to implications of narrow band spectra. Little attention to cross-correlation.

1.2   Robson J.D.: Introduction to Random Vibration: Edinburgh University Press 1963.
Elementary account of the subject, but with different emphasis from 1.1 Much more attention to cross-correlation and to general problem.

1.3   Crandall S.H.(ed): Random Vibration: Technology Press/Wiley 1958.
Essentially lecture notes of course given at MIT in 1957, giving good picture of the state of the subject at that time. Bibliography also helpful.

1.4   Crandall S.H.(ed): Random Vibration Vol.2: MIT Press 1963.
Similar course given in 1962 giving updated picture of state of the subject. Good chapters on non-linear and non-stationary problems, and on practical applications.

1.5   Bendat J.S.: Principles and Applications of Random Noise Theory: Wiley 1958.
Extended account of random noise theory including basic principles, not confined to problems of vibration.

1.6   Bendat J.S and Piersol A.G.: <u>Measurement and Analysis of</u>
         <u>Random Data</u>: Wiley 1965. Practically oriented
         treatment of theory based on experience of au-
         thor's "Measurement Analysis Corporation", large<u>_</u>
         ly in USA rocket program. Good account of theory,
         and its application to measurement of real random
         vibration.

1.7   Lin Y.K.: <u>Probabilistic Theory of Structural Dynamics</u>:
         McGraw-Hill 1967.
         Idiosyncratic but evidently sound development of
         random vibration theory from first principles to
         advanced problems. Tendency to enlarge on subjects
         which particularly interest author, but this is
         not necessarily a fault.

Chapter 2   DESCRIPTION I :   PROBABILITY

2.1. Probability

Suppose that we perform an experiment a number
of times under similar conditions and record the result in each
case. If we express the number of occasions on which a certain
result R occurs as a fraction of the total number of repetitions,
we obtain the 'relative frequency of the result R. The table
shows what happened in a particular experiment.

No. of repetitions:

50	100	150	200	250	300	350	400	450	500

No. of results R:

15	33	53	69	93	117	142	165	180	201

Relative frequency:

.30	.33	.353	.345	.378	.390	.406	.412	.400	.402

Although the relative frequency fluctuates, it
does tend to settle down within a diminishing band of values;
it exhibits 'statistical regularity'. If we repeated the whole
process, we would expect to get an ultimate value not far away
from 0.4. To obtain a mathematical model of this situation, we
assign to the result R a number p to correspond to our estimate
of the ultimate relative frequency, which we call the probability of R. In this case we could say that the probability was
0.4, on the basis of the tests reported above.

The relative frequency, and therefore the probability, must be a positive number in the range 0 to 1. We assign a probability of 1 to a result which is 'reasonably certain', and a probability of 0 to one which is 'highly unlikely'. If we consider the various alternative outcomes of an individual experiment, we can assign a probability to each outcome; and if these are mutually exclusive while covering all possibilities, one of these outcomes must occur and the total of the probabilities assigned must be unity. It is convenient to think that we have a unit of probability at our disposal which we have to apportion between the alternative results.

2.2. Continuous probability distribution

If the result of the hypothetical experiment is a value of a continuous variable x, which may take any value over a range, we work with a 'probability density function' $p(x)$, as a function of x over its whole range, thinking of our unit of probability as distributed over the range. This probability density is related to the lumped probability of the discrete case in the same way as a distributed load on a beam is related to a group of concentrated loads. In order that the total probability will be unity, we must have

$$\int_{-\infty}^{\infty} p(x)\, dx = 1 .$$

The probability of the occurrence of a value of x in the range $a < x < b$ may be written:

$$Pr\left[a < x < b\right] = \int_a^b p(x)dx$$

and for an elementary interval δx

$$Pr\left[x_0 < x < x_0 + \delta x_0\right] = p(x_0)\delta x_0 .$$

The units of $p(x)$ are 'probability (non–dimensional) per unit of x'. The probability of a value of x not exceeding x_0 is the 'probability distribution function' $P(x_0)$.

$$P(x_0) = Pr\left[-\infty < x \leqslant x_0\right] = \int_{-\infty}^{x_0} p(x)dx$$

and $p(x)$ is the derivation of $P(x)$ with respect to x. Evidently $P(x)$ increases from 0 to 1 as x increases from $-\infty$ to $+\infty$. We can use either $P(x)$ or $p(x)$ to define a particular distribution, whichever is more convenient. Typical forms for these functions are shown in Fig. 2.1.

2.3. Expectation

If we take each possible value of x, multiply by the probability of its occurrence, and form the sum, we get the 'expectation' of x. For a continuous variable x we write:

$$E\left[x\right] = \int_{-\infty}^{\infty} xp(x)dx .$$

This is obviously equal to the mean value of x which we should expect to obtain over a sufficiently large number of experiments.

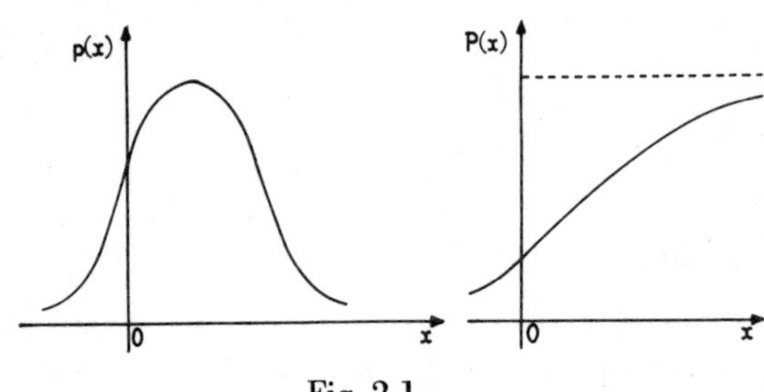

Fig. 2.1.

We can define similarly the expectation of any power of x , or of any function of x . In particular:

$$E[x^2] = \int_{-\infty}^{\infty} x^2 p(x)\,dx$$

is the mean square value of x, and $E[x^n]$ is termed the moment of the n'th order.

The values of the expectations depend solely on the form of the density function $p(x)$, although different functions could lead to the same values for some of the moments.

If the mean value $E[x]$ is $\bar{x}$, we can define similar 'central moments' in terms of the difference $(x-\bar{x})$. Then $E[x-\bar{x}]$ is of course 0.

$$E[(x-\bar{x})^2] \;=\; \int_{-\infty}^{\infty}(x-\bar{x})^2 p(x)\,dx \;=\; E[x^2]-(\bar{x})^2 \;=\; \sigma^2.$$

The quantity σ^2 is the 'variance' and σ itself the 'deviation'. If we can arrange for $\bar{x}$ to be 0, $E[(x-\bar{x})^2]$ becomes $E[x^2]$ and σ is the root mean square value.

We shall in fact be mainly concerned with second order quantities of the type of $E[x^2]$.

2.4. Types of probability density functions

These are to be chosen to match experimental rec̲ords of relative frequency in any particular case; but it is convenient to have available mathematical forms to suit commonly occurring types, with one or more parameters which can be adjusted to suit the circumstances. One such type has the following characteristics:

(1) a single peak at the most commonly occurring values of x,

(2) relative frequency decreasing symmetrically on either side of the peak.

Clearly the central value will be the mean value of x and we can make this correspond to $x = 0$, to get the form of Fig. 2.2.

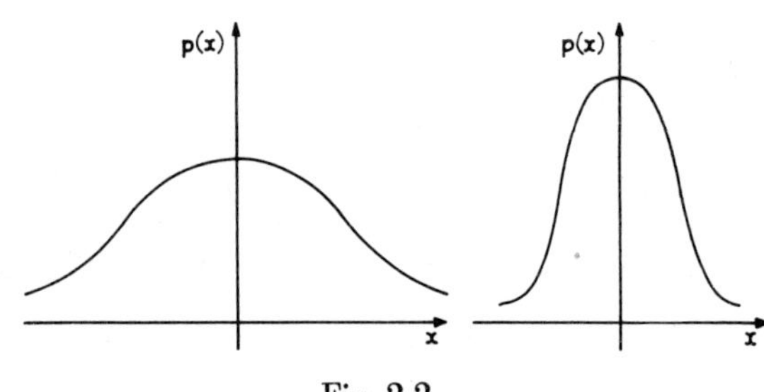

Fig. 2.2.

A convenient formula to represent such a probability density is of the type e^{-x^2}. This gives Gauss's normal distribution.

2.5. Gaussian density function

Assume $p(x) = Ae^{-c^2 x^2}$, with A and c constants, and mean value of $x = 0$.

We must have $\int_{-\infty}^{\infty} p(x)dx = 1$, which gives $\dfrac{A\sqrt{\pi}}{c} = 1$.

So $p = \dfrac{c}{\sqrt{\pi}} e^{-c^2 x^2}$.

The mean square value $\sigma^2 = \displaystyle\int_{-\infty}^{\infty} x^2 p(x)\,dx = \dfrac{1}{2c^2}$

so $p = \dfrac{1}{\sigma\sqrt{2\pi}} e^{-\frac{x^2}{2\sigma^2}}$, with the deviation σ as a single parameter.

Alteration of the value of σ affects the sharpness of the peak; we may note that the maximum value of p at $x = 0$ is $p_m = 1/\sigma\sqrt{2\pi}$, so that $\sigma = 0.4/p_m$, and also that the points of inflexion in the curve occur at $x = +\sigma$ and $-\sigma$.

The corresponding probability distribution function is

$$P(x) = \frac{1}{\sigma\sqrt{2\pi}} \int_{-\infty}^{x} e^{-\frac{x^2}{2\sigma^2}}\,dx \ .$$

The probability of a value $|x| < |x_0|$ is $2\left[P(x_0) - \dfrac{1}{2}\right]$.

If we use this form of density function, we require only one empirical parameter to define it completely, namely the value of $\sigma^2 = E\left[x^2\right]$, if the mean value of x is made 0.

It can be shown that this form of density function would be approached if the value of x was the sum of a large number of small contributions, each equally likely to be positive or negative in random manner. The diagram of Fig. 2.3 shows the result of 16 such random contributions, corresponding to a binomial distribution, compared with a Gaussian density function chosen to match. It is therefore reasonable to expect that this form will occur in random processes.

The special property of the mathematical form chosen for the Gaussian distribution, which distinguishes it from others, is that it retains its characteristic form under any linear transformation. It is this property that enables us to determine the effects of random processes, at least to a first approximation, in a relatively simple form.

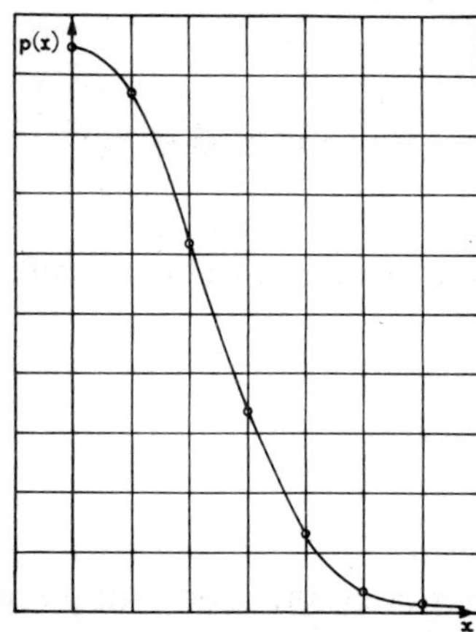

Fig. 2.3.

2.6. Tabulation

The integral defining $P(x)$ cannot be expressed in terms of elementary functions, but numerical values are given in most books on probability and statistics, and in books of tables of mathematical functions. Care is necessary in using these, since the functions tabulated are not at all the same; some varieties are listed below.

$$f = \frac{1}{\sqrt{2\pi}} e^{-\frac{x^2}{2}} \text{ , that is } p(x) \text{ with } \sigma = 1$$

$$f = \frac{1}{\sqrt{2\pi}} \int_{-\infty}^{x} e^{-\frac{u^2}{2}} du \text{ , that is } P(x) \text{ with } \sigma = 1$$

$$f = \frac{2}{\sqrt{\pi}} e^{-x^2} , \quad \text{that is } 2p(x) \quad \text{with } \sigma = \frac{1}{\sqrt{2}}$$

$$f = \frac{2}{\sqrt{\pi}} \int_0^x e^{-u^2} du , \quad \text{that is } 2\left[P(x) - \frac{1}{2}\right] \text{ with } \sigma = \frac{1}{\sqrt{2}}$$

This last quantity is called the 'error function', written erf x. The notation erfc x means $(1 - \text{erf } x)$.

2.7. Probability in relation to random processes

Consider any typical random process, and suppose that we make a number of records of the variation of some random quantity x as a function of time t (or of some other convenient independent variable). Each individual record is called a reali_ zation of the random process, which we can denote by $x_1(t), x_2(t)$, and so on. The records will differ in detail, but will have important characteristics in common, and we wish to express this situation in mathematical terms.

Considering a particular point in time t_0, we have a set of values of $x(t_0)$, namely $x_1(t_0)$, $x_2(t_0)$ and so on. We now assume that these values have a definite probability density function $p(x, t_0)$, corresponding to the relative frequency of occurrence of different values of x; this function will in general be different at different time instants so we include t_0 as a parameter.

Thus $p(x_A, t_0)\delta x_A = \Pr[x_A < x < x_A + \delta x_A]$ at $t = t_0$, where x_A is any particular value of x.

This would not commit us to any restriction on the relative values of x at successive times; but this is important, and we must introduce a probability density function to define a relation between the values of x (in any one realisation) at times t_1 and t_2 . We define a joint probability $p(x(t_1), x(t_2))$, such that

$$p(x_A, x_B)\,\delta x_A\,\delta x_B = Pr\left[x_A < x(t_1) < x_A + \delta x_A;\ x_B < x(t_2) < x_B + \delta x_B\right]$$

in any one realisation.

We could extend this to higher order joint probabilities, linking values at three or more instants, but for most purposes we take account of the first two orders only.

Mathematically we postulate that, although the course of the individual realisations is unknown, the various probability functions are specified, or at least capable of specification.

2.8. Stationary random process

We have left open the possibility of the probability density functions varying with time, but in many cases they will not vary over reasonable lengths of time. With this restriction, the random process is stationary; the first order probability will not then include the parameter t_0 , and the second order probability will not include the times t_1 and t_2 explicitly; it will however, depend on the interval between them

$$\tau = t_2 - t_1 .$$

We can see that the average value of x (over the realisations at any fixed time) is the expectation of x:

$$E[x] = \int_{-\infty}^{\infty} x p(x) dx, \quad \text{independent of } t .$$

In most cases the average value of a strictly random process can be made zero, by deducting if necessary a constant value which can be treated separately. With this simplification,

$$\sigma^2 = E[x^2] = \int_{-\infty}^{\infty} x^2 p(x) dx , \quad \text{also independent of } t .$$

The relation between successive values x_A and x_B, separated by a time interval τ is now defined by the mean value of the product $x_A x_B$.

$$\text{Thus} \quad E[x_A x_B] = \int\!\!\int_{-\infty}^{+\infty} x_A x_B p(x_A, x_B, \tau) dx_A dx_B ,$$

which is a function of the interval τ . This is the auto-correlation function which we denote by $R(\tau)$.

2.9. Ergodic random process

If a random process is stationary, it seems reasonable to expect that any one realisation will display all the features characteristic of the process if we continue to record it for a long enough range of time; if that is so, we should expect that the probability functions worked out from one realisation on a time basis would be the same as the corresponding quantities obtained by the previous method of studying the different realisations at specific times. The term 'ergodic' is applied to a random process which has this property. This is reasonably true of many processes, and we shall therefore introduce this restriction since it greatly simplifies the procedure. For this purpose we require theoretically an infinite record on a time basis; practically, we require a length of record sufficient to meet the condition above, namely to show all the characteristic features of the process.

2.10. Time averages for ergodic process

We use the notation: $<x(t)>$ for the time-average of x, and we can write:

$$E[x] \;=\; <x(t)> \;=\; \lim_{T \to \infty} \frac{1}{T} \int_{-\frac{1}{2}T}^{\frac{1}{2}T} x(t)\,dt$$

$$E[x^2] \;=\; <x(t)^2> \;=\; \lim_{T \to \infty} \frac{1}{T} \int_{-\frac{1}{2}T}^{\frac{1}{2}T} x(t)^2\,dt \;.$$

If the probability distribution over the realisations is Gaussian, these two quantities define the distribution. As noted before, we usually arrange for $<x>$ to be 0.

The mean value of the product of two values of

x, separated by an interval τ becomes:

$$R(\tau) = E\left[x_A x_B\right] = \,<x(t)x(t + \tau)> \,= \lim_{T \to \infty} \frac{1}{T} \int_{-\frac{1}{2}T}^{\frac{1}{2}T} x(t)x(t + \tau)dt .$$

2.11. Autocorrelation function

This function $R(\tau)$ plays an important part, and we can note at this stage some of its properties.

(1) From the integral, $R(\tau)$ is an even function of τ , that is $R(-\tau) = R(\tau)$.

(2) If $\tau = 0, R(0)$ becomes $<x^2>$ which is equal to σ^2 if $<x> = 0$. It is sometimes convenient to express $R(\tau)$ in a non-dimensional form by dividing by $R(0)$.

(3) In any random process, as τ increases, the link between the magnitude of $x(t + \tau)$ and $x(t)$ diminishes; ultimately the two values are practically unrelated and the average value of the product tends to zero, since products will be equally often positive and negative. Hence $R(\tau)$ tends to 0 as τ becomes large, and in fact $R(\tau)$ is always less than $R(0)$. (This applies to a strictly random process; if there is a periodic term present in x it will show up in R).

Typical standard forms for $R(\tau)$ which occur in theoretical work are shown in Fig. 2.4. These ca be represented by relations of the types:

$$R(\tau) = a\,e^{-b|\tau|} \quad \text{and} \quad a\,e^{-b|\tau|}\cos c\tau\,,$$

where **a**, **b**, and **c** are constants to be adjusted to suit the circumstances.

If we wish to consider the function $x(t)$ as the sum of two functions, say $u(t)$ and $v(t)$, representing perhaps a random and a determinate function, the auto-correlation function becomes:

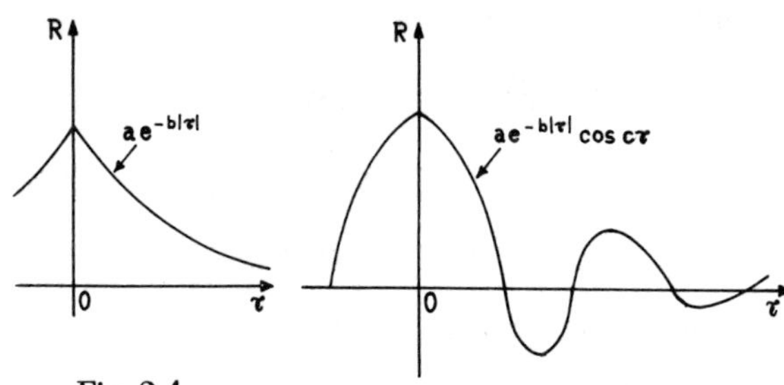

Fig. 2.4.

$$R_x(\tau) = \lim_{T \to \infty} \frac{1}{T} \int_{-\frac{T}{2}}^{\frac{T}{2}} \left\{ u(t)u(t+\tau) + v(t)v(t+\tau) + u(t)v(t+\tau) + v(t)u(t+\tau) \right\} dt$$

$$= R_u(\tau) + R_v(\tau) + R_{uv}(\tau) + R_{vu}(\tau)\,,$$

where the first two terms are auto-correlation functions for u and v, and the remaining terms are 'cross-correlation functions' measuring the link between u and v. These will be discussed later; meanwhile we note that they will be zero only if u and v are completely independent (as they will be if one is random and the other determinate).

Bibliography for Chapter 2

All the books listed in the Chapter 1 contain some account of probability theory, and those by Bendat and Lin have quite a lot. To present the theory on a logical mathematical basis, which will be valid for discrete and for continuous distributions, requires a fairly sophisticated treatment, and this is adopted in most modern books on the subject. Of those given below, Parzen's treatment is at a medium level. Cramér is more advanced.

2.1. Parzen E. Modern probability theory and its applications. Wiley 1960.

2.2. Parzen E.: Stochastic Processes. Holden–Day 1962. (both these are available as paperbacks).

2.3. Cramér H.: Mathematical methods of statistics. Princeton University Press 1958.

Chapter 3   DESCRIPTION II : SPECTRAL DENSITY

3.1. Frequency analysis

The probability functions which we have considered give an account of the part played in a random process by values of x of different magnitudes, but they do not reveal the course of the variation of x with time except through the auto-correlation function $R(\tau)$ as a function of time interval τ . We are accustomed in vibration work to think of time variation as measured rather by frequency, and we now consider how we can apply this concept to a random function, through the method of Fourier analysis. We shall find that this leads us back to the auto-correlation function by a different path.

3.2. Fourier series

We shall state briefly some results which we wish to use. If we have a function $x(t)$ of t, periodic with a period T, we can express x as the sum of an infinite series of sines and cosines of integral multiples of the basic frequency $\omega = 2\pi/T$.

Thus
$$x(t) = a_0 + a_1 \cos \omega t + a_2 \cos 2\omega t + \ldots$$
$$+ b_1 \sin \omega t + b_2 \sin 2\omega t + \ldots$$

in which the a's and b's are constant coefficients given by

the formulae:

$$a_n = \frac{2}{T} \int_{-\frac{1}{2}T}^{\frac{1}{2}T} x(t)\cos n\omega t\, dt \quad \text{and} \quad b_n = \frac{2}{T} \int_{-\frac{1}{2}T}^{\frac{1}{2}T} x(t)\sin n\omega t\, dt$$

and $\quad a_0 = \dfrac{1}{T} \displaystyle\int_{-\frac{1}{2}T}^{\frac{1}{2}T} x(t)\, dt \cdot n$ represents any positive integer.

These integrals may extend over any complete period of x.
The restrictions on the form of the function x, and the sense
in which the series can replace the function, are discussed in
books on Fourier series, but they need not detain us here. It is
sufficient to note that $x(t)$ need not be smooth or continuous.

The time average of x can be expressed in terms
of the coefficients in the series:

thus $\qquad \langle x(t)\rangle = a_0$

$$\langle x^2(t)\rangle = \frac{1}{T} \int_{-\frac{1}{2}T}^{\frac{1}{2}T} \big[x(t)\big]^2 dt = a_0^2 + \frac{1}{2}\sum_1^\infty (a_n^2 + b_n^2) \, .$$

For our purpose it is more convenient to express these results
in terms of complex exponentials. Thus

$$a_n \cos n\omega t + b_n \sin n\omega t = c_n e^{in\omega t} + c_{-n} e^{-in\omega t} \, ;$$

a_0 will be c_0, with $c_n = \frac{1}{2}(a_n - ib_n)$ and $c_{-n} = \frac{1}{2}(a_n + ib_n)$,
conjugate numbers. We can write: $x(t) = \displaystyle\sum_{-\infty}^{\infty} c_n e^{in\omega t}$, with n
taking all integral values, positive and negative, and zero.

The coefficients expressed as integrals are now:

$$c_n = \frac{1}{T} \int_{-\frac{1}{2}T}^{\frac{1}{2}T} x(t)\, e^{-in\omega t}\, dt \ .$$

We now have: $\langle x(t) \rangle = c_0$

$$\langle x^2(t) \rangle = \sum_{-\infty}^{\infty} |c_n|^2 \ .$$

$|c_n|$ is the modulus of the complex number c_n, so $|c_n|^2 = |c_{-n}|^2 =$
$= \frac{1}{4}(a_n^2 + b_n^2)$.

3.3. Fourier integral

The results quoted are valid for periodic functions, and they can be applied to a finite stretch of a non-periodic function by treating this stretch as one cycle of a hypothetical periodic function. To apply them to the complete range
of a non-periodic function, we require to extend the period to
infinity and to reduce the basic frequency to zero. This leads
to the concept of a Fourier integral, replacing the sum of terms
in the series. Write the series in the form:

$$x(t) = \sum_{-\infty}^{\infty} c_k\, e^{ik\omega_0 t} \quad \text{and} \quad c_k = \frac{1}{T} \int_{-\frac{1}{2}T}^{\frac{1}{2}T} x(t)\, e^{-ik\omega_0 t}\, dt$$

where we have used ω_0 for the fixed basic frequency. We wish to
express $x(t)$ in terms of a continuous frequency ω in a form:

$$x(t) \;=\; \frac{1}{2\pi} \int_{-\infty}^{\infty} A(i\omega)e^{i\omega t}\cdot d\omega \;, \text{ with } \;\; A(i\omega) \;\; \text{a complex}$$

function of ω. (The factor 2π will enter the calculation and it is convenient to place it here). The product $\dfrac{A(i\omega)d\omega}{2\pi}$ corresponds to c_k in the pervious series, so that the dimensions of A are different from c. We therefore put $c_k \;=\; \dfrac{A_k \omega_0}{2\pi}$; then

$$x(t) \;=\; \frac{1}{2\pi} \sum A_k \omega_0 e^{ik\omega_0 t} \;\; \text{and} \;\; A_k \;=\; \int x(t)e^{-ik\omega_0 t}\cdot dt$$

since $\omega_0 T = 2\pi$. Now put $\;\; \omega_n = k\omega_0,$ so that $\omega_0 = \omega_{n+1} - \omega_n = \delta\omega$; A_k is now $A(\omega_n)$; then

$$x(t) \;=\; \frac{1}{2\pi} \sum A(\omega_n)\cdot e^{i\omega_n t} \;\;\; \text{and} \;\; A(\omega_n) \;=\; \int x(t)\cdot e^{-i\omega_n t}\, dt \;.$$

If we take the limiting form of this for $\omega_0 = \delta\omega \to 0$ and $T \to \infty$

$$x(t) \;=\; \frac{1}{2\pi} \int_{-\infty}^{+\infty} A(i\omega)\cdot e^{i\omega t}\cdot d\omega \;\; \text{and} \;\; A(i\omega) \;=\; \int_{-\infty}^{\infty} x(t)\cdot e^{-i\omega t}\cdot dt \;.$$

The first of these relations is a Fourier integral, and $A(i\omega)$ is the Fourier transform of $x(t)$. $A(i\omega)$ is a complex function of ω as we can see by writing it in trigonometric terms:

$$A(i\omega) \;=\; \int_{-\infty}^{\infty} x(t)\cdot \cos\omega t\cdot dt \;-\; i\int_{-\infty}^{\infty} x(t)\cdot \sin\omega t\cdot dt \;=\; \alpha(\omega) - i\beta(\omega) \;,$$

where α and β are real functions of ω , α being an even function and β odd. Evidently these integrals will not exist unless $x(t) \to 0$ for $t \to \infty$ and $-\infty$.

In the nearly-symmetrical relations between $x(t)$

and $A(i\omega)$, the factor 2π must occur somewhere, and the notation will vary according to where this factor is placed. For our purpose, it is convenient to eliminate it by expressing the formulae in terms of the frequency f in place of the circular frequency (pulsatance) ω, with $\omega = 2\pi f$. This gives:

$$x(t) = \int_{-\infty}^{\infty} A(if) \cdot e^{i2\pi ft} \cdot df \quad \text{and} \quad A(if) = \int_{-\infty}^{\infty} x(t) \cdot e^{-i2\pi ft} dt .$$

3.4. Mean values in terms of Fourier transfrom and spectral density

We cannot obtain directly mean values over an infinite range; to overcome the difficulty we introduce a modified function $x_\tau(t)$ such that $x_\tau(t) = x(t)$ over the range $-\frac{1}{2}T$ to $+\frac{1}{2}T$, and $x_\tau(t) = 0$ outside this range. Having defined the required quantity in terms of $x_\tau(t)$, we take the limit for $T \to \infty$. Thus the mean value of $x_\tau(t)$ over the range T is

$$\frac{1}{T} \int_{-\frac{1}{2}T}^{\frac{1}{2}T} x_\tau(t) \, dt = \frac{1}{T} \int_{-\infty}^{\infty} x_\tau(t) \, dt = \frac{1}{T} A_\tau(0)$$

with $\quad A_\tau(if) = \displaystyle\int_{-\infty}^{\infty} x_\tau(t) e^{-i2\pi ft} \cdot dt \quad$ and $\quad <x(t)> = \lim_{T \to \infty} \left[\frac{1}{T} A_\tau(0) \right] .$

For the mean square value, we have in our previous notation

$$\int_{-\frac{1}{2}T}^{\frac{1}{2}T} x^2(t)\,dt \;=\; T\sum_{-\infty}^{\infty}|c_k|^2 \;=\; T\sum\left|\frac{A_k\omega_0}{2\pi}\right|^2 \quad \text{with}\quad \omega_0 T \;=\; 2\pi$$

$$\therefore \int_{-\infty}^{\infty} x^2(t)\,dt \;=\; \frac{1}{2\pi}\int_{-\infty}^{\infty}|A(i\omega)|^2\cdot d\omega\ .$$

Here $|A(i\omega)|^2 = [\alpha(\omega)]^2 + [\beta(\omega)]^2$, a real even function of ω ; so we can take the integral over positive frequencies only

$$\int_{-\infty}^{\infty} x^2(t)\,dt \;=\; \frac{1}{\pi}\int_{0}^{\infty}|A(i\omega)|^2\,d\omega \;=\; 2\int_{0}^{\infty}|A(if)|^2\,df\ .$$

Now introducing $x_\tau(t)$, mean value over range T is

$$\frac{1}{T}\int_{-\frac{1}{2}T}^{\frac{1}{2}T} x_\tau^2(t)\cdot dt \;=\; \frac{1}{T}\int_{-\infty}^{\infty} x_\tau^2(t)\cdot dt \;=\; \frac{2}{T}\int_{0}^{\infty}|A_\tau(if)|^2\cdot df$$

$$\therefore\ <x^2(t)> \;=\; \int_{0}^{\infty}\left\{\lim_{T\to\infty}\frac{2}{T}|A_\tau(if)|^2\right\}df \;=\; \int_{0}^{\infty} S(f)\cdot df$$

$$\text{with}\ \ S(f) = \lim_{T\to\infty}\left[\frac{2}{T}|A_\tau(if)|^2\right] \ \ \text{and}\ \ A_\tau(if) = \int_{-\infty}^{\infty} x_\tau(t)\cdot e^{-i2\pi ft}\cdot dt\ .$$

This function S of the frequency is the 'spectral density', which proves to be the key to our treatment of random functions.

3.5. Spectral density

From the relation $\langle x^2(t) \rangle = \int_0^\infty S(f)\,df$ we can see the physical meaning of this quantity, namely that $S(f)\,df$ for any range of frequency from f to $(f + df)$ measures the contribution to the mean square of $x(t)$ coming from this frequency range. Since in many applications the quantity $\langle x^2(t) \rangle$ measures the mean power, S is sometimes called the 'power spectral density'. Experimentally we should have to define df by a narrow-band filter and measure the value of x^2 transmitted.

From the equations it will be seen that $S(f)$ is a real, even function of f. The notation used for S varies according as the integral is taken one-sided (as above) or double-sided, which introduces a factor of 2, and also according as S is considered as a function of f or of ω, which introduces a factor of 2π. The convention used here is that used in the books by Bendat and Robson in the bibliography to Chapter 1.

Bibliography for Chapter 3

Fourier series and integrals are discussed in many books on mathematics. The two listed below are, however, written with engineering applications in view.

3.1. Lawden D.F.: <u>Mathematics of Engineering Systems</u>:
Methuen 1959 (also paperback).
This applies the theory to random processes.

3.2. Kaplan W.: Operational Methods for Linear Systems:
Addison–Wesley 1962.
More advanced in treatment, but does not touch random processes.

Chapter 4   VIBRATION THEORY I : RECEPTANCE

4.1. Forced vibration of a damped system having one degree of freedom

When the configuration of a mechanical system is completley described by one independent variable x, it is said to have one degree of freedom. This quantity usually has the dimension of length or angle. In a real system there will be inertia forces depending on the second time derivative $\ddot{x}$, viscous damping forces depending on the velocity $\dot{x}$, and restoring forces depending on x. The equation governing the vibration can usually be brought into the form

$$f(\ddot{x},\dot{x},x) \;=\; P(t) , \tag{1}$$

where the periodic excitation $P(t)$ is regarded as the input to the system and the resulting motion $x(t)$ the response. It should be noted in passing that in requiring the forcing term $P(t)$ to appear alone in (1) we have ruled out the case of parametric excitation. The system is said to be linear when equation (1) can be written in the standard form

$$\ddot{x} + 2\zeta\omega_0\dot{x} + \omega_0^2 x \;=\; P(t) , \tag{2}$$

i.e. when no higher power of $x,\dot{x},\ddot{x}$ occurs. (The system is still linear when either of the two parameters in (2) varies with time.

However, we will exclude such cases here).

Above we have had in mind a force as input and a displacement output. There are cases where the excitation is a prescribed displacement causing another displacement as response. There are also cases where the governing equation has to be formulated in order to find out what the input to the system really is. For instance, the casing shown below contains a plunger m constrained by spring k and dashpot c.

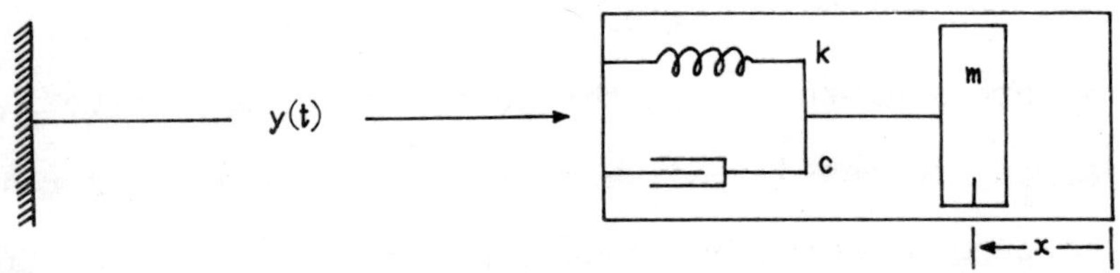

The prescribed (or unknown) displacement $y(t)$ excites the system. However, since the governing equation is $m(x-y)^{\cdot\cdot} + c\dot{x} + kx = 0$, or

$$(3) \qquad \ddot{x} + \frac{c}{m}\dot{x} + \frac{k}{m}x = \ddot{y}(t)$$

we must take $P(t) = \ddot{y}$ as the input. For this reason the system is known as an accelerometer.

Comparison of (2) and (3) yields the damping factor $\zeta = c/(2\sqrt{km})$ and the undamped natural frequency $\omega_0 = \sqrt{k/m}$ rad/sec. The corresponding cyclic frequency is

$$f_0 = \omega_0/2\pi \qquad \text{hertz (or cycles per second).}$$

There are several ways of solving equation (2).

In vibration theory we study stable systems, so any transients will be damped out with time. Thus we search for the steady-state solution only. When $P(t) = P_0 m^{-1} \cos 2\pi f t = P_0 m^{-1} \cos \omega t$, it is not too surprising that the calculations can be simplified by using the identity

$$e^{i 2\pi f t} \;=\; \cos 2\pi f t + i \sin 2\pi f t \,,$$

where $i^2 = -1$. The panalty is that one must deal with complex-valued quantities even though the original equation was a real one in real variables. This technique consists in studying not the original equation (2) but the analogous equation

$$\underline{\ddot{x}} + 2\zeta \omega_0 \underline{\dot{x}} + \omega_0^2 \underline{x} = P_0 m^{-1} e^{i 2\pi f t} \,, \tag{2}'$$

in which $\underline{x} = x + iy$ is the complex response. Then, provided $\mathrm{Re}(\underline{\dot{x}}) = (\mathrm{Re}(\underline{x}))^{\textstyle\cdot} = \dot{x}$ and $\mathrm{Re}(\underline{\ddot{x}}) = (\mathrm{Re}(\underline{x}))^{\cdot\cdot} = \ddot{x}$, x will be the solution of (2) with $P(t) = P_0 m^{-1} \cos 2\pi f t$.

4.2. Receptance $\alpha(if)$

The receptance $\alpha(if)$ of the system is defined by writing the complex response in the form

$$\underline{x}(t) \;=\; \alpha(if) P_0 \, e^{i 2\pi f t} \,, \tag{4}$$

Much confusion can be caused by the common prac tice of not distinguishing between the actual response x and the complex response $\underline{x}$. We will emphasize the fact that the receptance is complex by regarding its argument not as f but as if. The magnitude of the receptance $|\alpha(if)|$ is the ratio of the response amplitude at each frequency. This is the gain $g(f)$ or $g(\omega)$ of the system. The phase shift θ between the response x and the input P is given by $\arg\alpha = -\theta(\omega)$. Thus, the performance of the single-freedom system is completely described once the real and <u>imagin</u>ary components of α are known at all frequencies. (The <u>receptance</u> is also known as the complex frequency response, and the <u>gain</u> as the magnification factor, frequency response, or transfer function).

Substitution of (4) into (2)' yields

$$\alpha(if) = \frac{(4\pi^2)^{-1}}{(f_0^2 - f^2 + 2i\zeta ff_0)m}$$

$$(5) \qquad = \frac{1}{(\omega_0^2 - \omega^2 + 2i\zeta\omega\omega_0)m}$$

$$= \frac{\omega_0^2 - \omega^2 - 2i\zeta\omega\omega_0}{\{(\omega_0^2 - \omega^2)^2 + 4\zeta^2\omega^2\omega_0^2\}m} = g(\omega)e^{-i\theta(\omega)} \ .$$

Hence, the gain $g(\omega)$ and phase shift $\theta(\omega)$ are

$$(6) \qquad g(\omega) = m^{-1}\{(\omega_0^2 - \omega^2)^2 + 4\zeta^2\omega^2\omega_0^2\}^{-\frac{1}{2}},$$

$$(7) \qquad \theta(\omega) = \arctan(2\zeta\omega\omega_0/(\omega_0^2 - \omega^2)).$$

These are indicated by the continuous curve and the dashed curve,

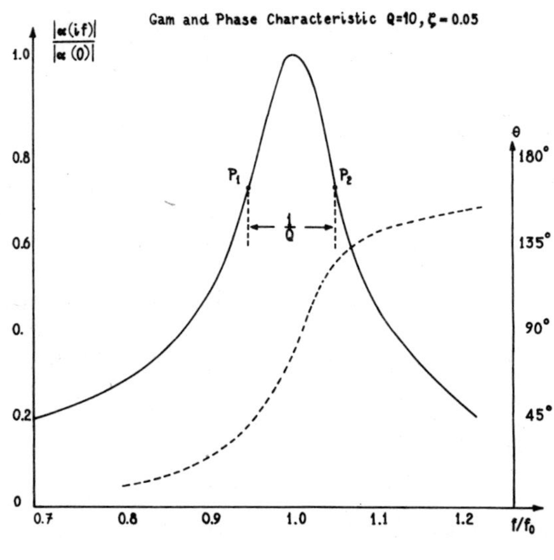

Fig. 4.1

respectively, in Fig. 4.1. Since the receptance is complex, it is able to describe both the gain and phase characteristic of the linear system.

4.3. Quality factor Q

First we note that the phase shift increases con<u>tin</u>ually with frequency from zero at zero frequency to its great<u>est</u> value of 180°. The phase shift is exactly 90° at the undamped natural freqency ω_0 ; from (7) $\tan\theta = \infty$ when $\omega = \omega_0$.

The **Q** factor is a measure of the resonant magnification of the system and can be defined as the ratio

$$Q = \frac{|\alpha(if_0)|}{|\alpha(0)|} = \frac{g(\omega_0)}{g(0)} . \tag{8}$$

It is to be noted in passing that the maximum gain actually occurs at the frequency $\omega = \omega_0(1 - 2\zeta^2)^{\frac{1}{2}}$. However,

when the viscous damping is small the difference is negligible,
and when the damping is large the hump in the gain curve is so
flat that the difference is again negligible. (In steel at engin_
eering strain levels, $\zeta \doteq 0.003$).

Definition (8) is not convenient for experimental
purposes because of the difficulties in measuring $g(0)$ at zero
frequency. Nevertheless, with the values $mg(\omega_0) = (2\zeta\omega_0^2)^{-1}$, $mg(0) = \omega_0^{-2}$
given by (6) we obtain the important theoretical relation

$$(9) \qquad\qquad Q = \frac{1}{2\zeta} .$$

Using the above value of the damping factor for steel we obtain
$Q = 160$. At lower strain levels the Q factor can be much higher.

A more commonly used relation defining Q is found
as follows: Let ω_1 and ω_2 be the frequencies at which the phase
leads or lags that at resonance by 45°, respectively. Using (7)
we have

$$2\zeta\omega_0\omega_1/(\omega_0^2 - \omega_1^2) = +1 , \quad 2\zeta\omega_0\omega_2/(\omega_0^2 - \omega_2^2) = -1 .$$

Hence

$$\omega_1 = -\zeta\omega_0 + (1 + \zeta^2)^{\frac{1}{2}}\omega_0 ,$$

and

$$\omega_2 = \zeta\omega_0 + (1 + \zeta^2)^{\frac{1}{2}}\omega_0 .$$

Thus

$$\omega_2 - \omega_1 = 2\zeta\omega_0 ,$$

or

$$\frac{\omega_0}{\omega_2 - \omega_1} = \frac{f_0}{f_2 - f_1} = \frac{1}{2\zeta} = Q . \tag{10}$$

It can be verified that

$$|\alpha(if_1)| = g(\omega_1) = 1/2m\sqrt{2}\,\zeta\omega_1\omega_0 ,$$

$$|\alpha(if_2)| = g(\omega_2) = 1/2m\sqrt{2}\,\zeta\omega_2\omega_0 ,$$

and we know that

$$|\alpha(if_0)| = g(\omega_0) = 1/2m\zeta\omega_0^2 .$$

Once again assuming that Q is fairly high, it can be said that the gain at ω_1 and ω_2 is $1/\sqrt{2}$ that at resonance. This attenuation in decibels is $20\log_{10}\sqrt{2} \doteq 3db$. Thus, these points, marked P_1 and P_2 on Fig. 4.1, are often called the three–decibel points. Since the power during steady state conditions in a lin ear system is proportional to the square of the amplitude, P_1 and P_2 are also called the half–power points.

Experimental data on the damping capacity of metals is often expressed not in terms of ζ but rather in terms of the logarithmic decrement δ or, more commonly, in terms of the specific damping capacity $\Delta W/W$, where ΔW is the energy lost per cycle and W is the maximum potential energy during the cycle. (Unfortunately there is a host of symbols used for these quantities). If A_n , A_{n+1} are the amplitudes of two successive positive peaks in the damped transient, then

$$\Delta W/W = (A_n^2 - A_{n+1}^2)/A_{n+1}^2 \doteq 2(A_n - A_{n+1})/A_{n+1} \, ,$$

and

$$\delta = \ell n(A_n/A_{n+1}) \doteq (A_n - A_{n+1})/A_n \, .$$

Moreover, the transient solution of (2) to be given in Lecture 5, shows that

$$A_n/A_{n+1} = \exp(2\pi\zeta(1 - \zeta^2)^{-\frac{1}{2}})$$

$$\doteq \exp(2\pi\zeta) \, .$$

Whence we have the conversion formulae

$$2\pi\zeta = \delta = \frac{1}{2}\frac{\delta W}{W} \, .$$

The data given in Fig. 4.2 is insufficient for

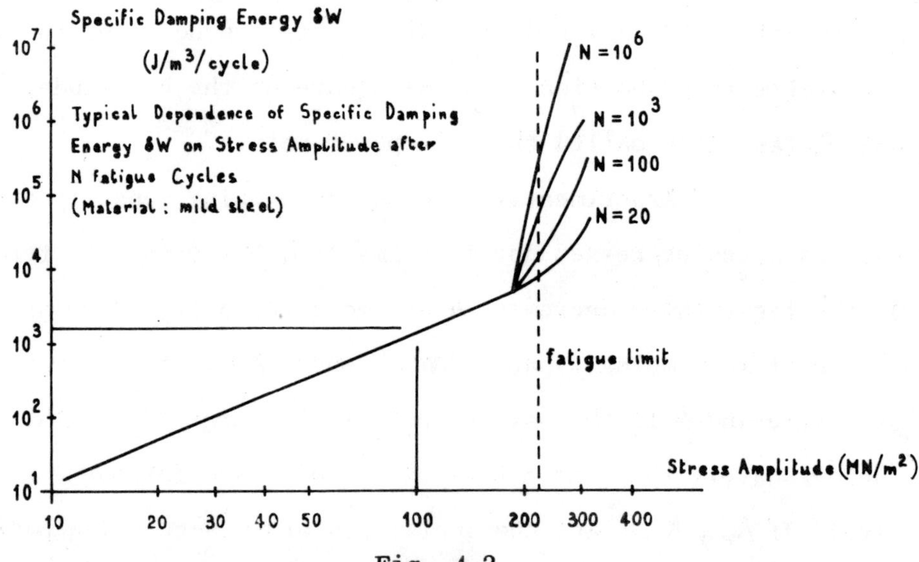

Fig. 4.2

the calculation of ζ because only δW is given as a function of stress level; to calculate W we need to know Young's modulus and set

$$W \;=\; \sigma^2(\text{peak})/2E \;=\; \sigma^2(\text{rms})/E \ .$$

However, if we take $E = 2\times10^{11}\,\text{N/m}^2$ for mild steel at $\sigma = 100\,\text{MN/m}^2$ we obtain $\delta W/W = 0.03$ and hence $\zeta = 0.002$.

4.4. Receptance in composite system

Consider the case of a damper on an elastic foundation. The equation for the spring is

$$\underline{x}_3 \;=\; \frac{1}{k}\underline{F} \ .$$

For the mass

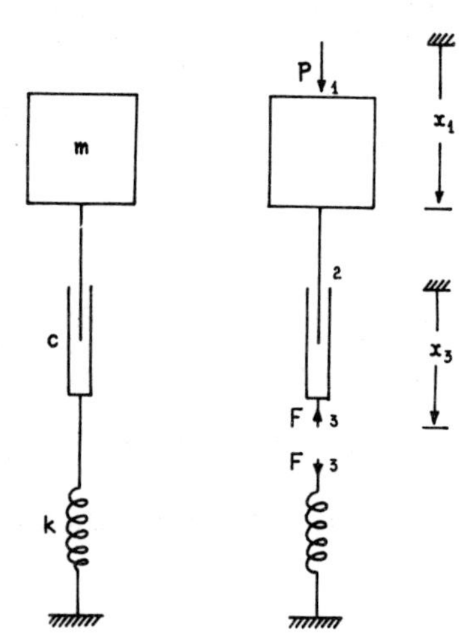

$$\underline{x}_1 \;=\; \alpha_{11}\underline{P} \;-\; \alpha_{12}\underline{F}$$

$$\;=\; \frac{-1}{m\omega^2}\underline{P} \;+\; \frac{1}{m\omega^2}\underline{F} \ .$$

For the dashpot

$$\underline{x}_3 \;=\; \underline{x}_1 \;-\; \frac{1}{i\omega c}\underline{F} \ .$$

Elimination of the internal force $\underline{F}$ and displacement $\underline{x}$, yields

$$\underline{x}_3\!\left(-m\omega^2 \;+\; k \;+\; im\frac{\omega}{c}k\right) \;=\; \underline{P}_1 \ .$$

However, by definition of α_{31}, we have

$$\underline{x}_3 = \alpha_{31}\underline{P}_1$$

so

$$\alpha_{31} = \frac{1}{m(\omega_0^2 - \omega^2 + i\omega k/c)} \ .$$

(The suffix notation is the same as for influence numbers. The first suffix gives the station where the response occurs and the second suffix gives the location of the excitation).

By applying a force $\underline{P}_3$ at station 3 and calculating the complex response $\underline{x}_1$ of the mass, we find from the relation

$$\underline{x}_1 = \alpha_{13}\underline{P}_3$$

that

$$\alpha_{31} = \alpha_{13} \ .$$

This is an important reciprocal property of receptances that holds even for continuous systems.

Chapter 5   **VIBRATION THEORY II: IMPULSIVE RECEPTANCE**

5.1. Introduction

In the preceding lecture the excitation $P(t)$ of a linear system was considered as a superposition of sinusoidal components. By virtue of the linearity of the system, each sinusoidal component can be treated separately and the corresponding responses added thus leading in a natural way to the concept of a frequency-dependent function, the receptance $\alpha(if)$, describing the behaviour of the system.

In the present lecture a time-wise decomposition of the excitation into a series of equivalent impulses will be considered. This leads to a time-dependent function, the so-called impulsive receptance $W(t)$ describing the behaviour of the system. It will be shown that for a given system $\alpha(if)$ and $W(t)$ are in general a Fourier transform pair, i.e., each is the Fourier transform of the other, thus establishing the theoretical equivalence of the two descriptions.

Each of the above two receptances will be used at times in the following lectures, and it will become evident that the choice of one or other receptance can depend on (1) analytical convenience, (2) whether interpretations in the time-domain or frequency-domain are easier in the physical problem in hand, (3) computational efficiency in the case of large-scale data handling.

5.2. Response to deterministic excitation P(t). Sampled data

It is becoming quite usual to record the values of a variable $P(t)$ only at uniformity spaced times $t = n\Delta$, where n are the integers and Δ is the sampling period. For the purpose of calculating responses it is then convenient to assume that $P(t) = P_n = P(n\Delta)$ during the sampling interval $(n-1)\Delta < t < n\Delta$ (See Fig. 5.1.) When the system is linear the effects of each constant loading P_n can be treated separately and the corresponding responses can be added at any subsequent time to give the total response of the system.

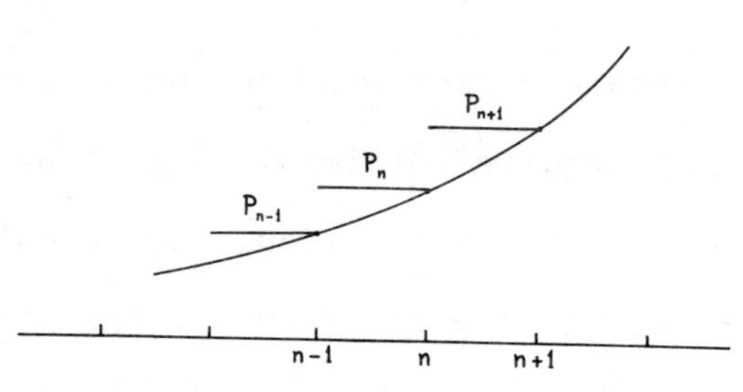

Fig. 5.1

The response at time $s\Delta$ after a loading of unit intensity for a duration Δ will be denoted by W_s, as shown in Fig. 5.2. The response at time $r\Delta$ will be denoted x_r. Thus, for example, the contribution to x_9 due to P_1 is $P_1 W_8$ and that due to P_5 is $P_5 W_4$, as indicated in Fig. 5.3. In general we have

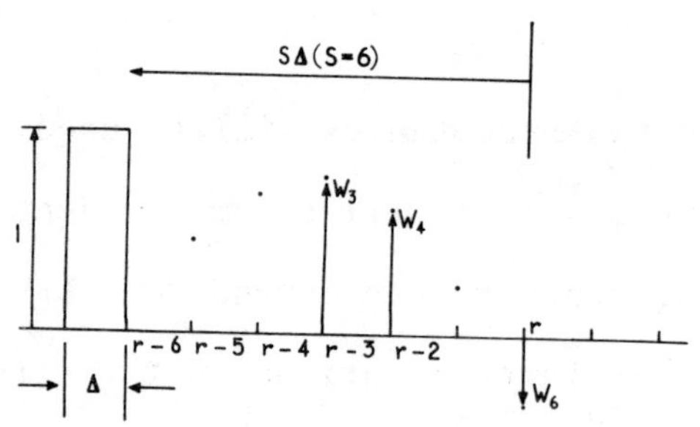

Fig. 5.2

$$x_r \; = \; P_{r-1}W_1 + P_{r-2}W_2 + P_{r-3}W_3 + \dots \; ,$$

$$= \sum_{s=1}^{\infty} P_{r-s}W_s \; . \tag{1}$$

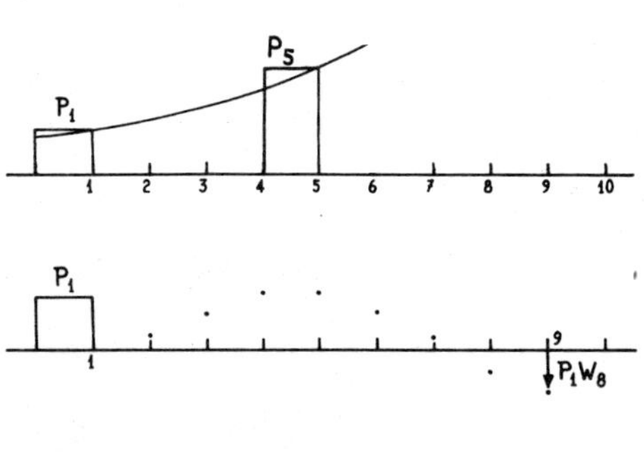

Fig. 5.3

If it is agreed that $W_s = 0$ for $s < 0$, (1) can be written in the form

$$x_r \; = \; \sum_{s=-\infty}^{+\infty} P_{r-s}W_s \; . \tag{2}$$

Change of the summation index yields the alternative formula

$$x_r \; = \; \sum_{t=-\infty}^{+\infty} P_t W_{r-t} \; . \tag{3}$$

It is worth noting here that the so-called convolutions (2), (3) have become very important in data processing, where the values W_s are called _weights_ and the convolution is usually called a _digital filter_. These matters will be referred to in the later lectures on data analysis.

5.3. Response to deterministic excitation. Continuous data

The appropriate expression for the continuous response $x(t)$ to a continuous excitation $P(t)$ will be derived from expression 5.2(1) by considering what happens in the limit as

$\Delta \longrightarrow 0$, i.e. as the sampling rate becomes infinite. (A device that converts a continuous "signal" $P(t)$ into a sampled time series P_n is called an analogue digital to converter, an <u>AD converter</u>, and the rate Δ^{-1} can usually be preselected in such a device).

The sampling period Δ will usually be shorter than half the shortest period of interest, and if Δ were further decreased it would be found that (for a linear system with some inertia) the response to the constant excitation $P = h$ tends rapid<u>l</u>y to a limiting shape $W(\tau) h \Delta$. Fig. 5.4. indicates the rapidity of the limiting process for a one freedom system. The area $h\Delta$ is thus a measure of the "impulse", and the <u>unit impulse function</u> $\delta(\tau)$ can be thought of as the limit of such an excitation as $\Delta \longrightarrow 0$ with $h\Delta = 1$.

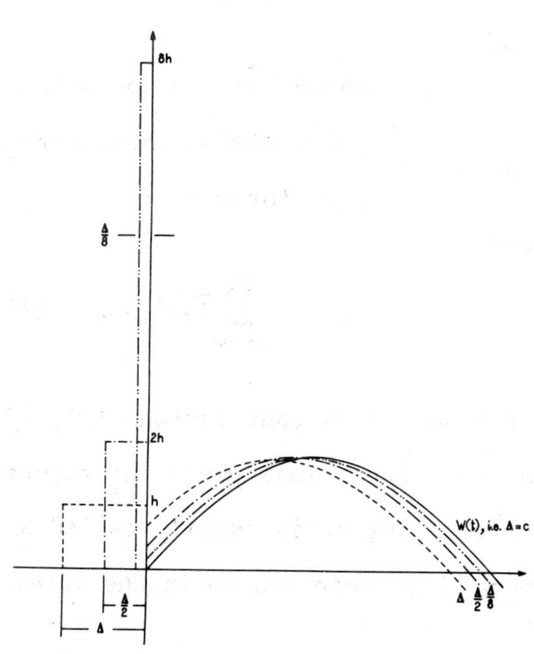

Fig. 5.4

If we now set $r\Delta = t$, $s\Delta = \tau$ and $x_r = x(r\Delta) = x(t)$, it follows that W_s approach $W(\tau)\Delta$ and in the limit the sum in 5.2(1) becomes

$$x(t) \;=\; \lim_{\Delta \to 0} \sum_{s=1}^{\infty} P(t-\tau)W(\tau)\Delta \;,$$

i.e.

$$x(t) \;=\; \int_0^{\infty} P(t-\tau)W(\tau)\,d\tau \;. \tag{4}$$

When dealing with Fourier transforms it can be more convenient to write

$$x(t) \;=\; \int_{-\infty}^{+\infty} P(t-\tau)W(\tau)\,d\tau \;, \tag{5}$$

remembering that $W(\tau)=0$ when $\tau<0$. The change of variable $t-\tau = s$ then yields

$$x(t) \;=\; \int_{-\infty}^{+\infty} P(s)W(t-s)\,ds \;, \tag{6}$$

which is known as Duhamel's integral when written as

$$x(t) \;=\; \int_{-\infty}^{t} P(s)W(t-s)\,ds \;. \tag{6a}$$

The function $W(\tau)$ is a receptance in the sense that it determines the response $x(t)$ to any given excitation $P(t)$ through a closed expression such as (5) or (6). Since $W(\tau)$ is in fact the response to a unit impulse excitation, as seen formally by setting $P(t-\tau)=\delta(t-\tau)$ in (5) and noting that $\int_{-\infty}^{+\infty} F(\tau)\delta(t-\tau)d\tau =$ $= F(t)$ for any continuous $F(\tau)$, we call $W(\tau)$ the <u>impulsive receptance</u>.

5.4. $W(\tau)$ and $\alpha(if)$ are Fourier transform pair

This relationship between the two receptances will be established for stable passive systems, i.e. systems in which a bounded excitation always causes a bounded response. According to 5.2.(4) the bounded excitation $P(-\tau)=\text{sign}\,W(\tau)$ yields the response $x(0) = \int_{0}^{\infty}|W(\tau)|d\tau$, which certainly will be bounded for a stable system. This means that $W(\tau)$ must attenuate to zero quite rapidly as τ increases. Moreover, it means that the integral

$$(7) \qquad \underline{x}(t) = \int_{-\infty}^{+\infty} W(\tau)P_0\, e^{i2\pi f(t-\tau)}d\tau$$

obtained from 5.2(5) by setting $P(s) = P_0 e^{i2\pi fs}$, as is done when determining the receptance, converges without any mathematical difficulties. If (7) is written in the form

$$(8) \qquad \underline{x}(t) = \left\{\int_{-\infty}^{+\infty} W(\tau)e^{-i2\pi f\tau}d\tau\right\}P_0\, e^{i2\pi ft}$$

and is compared with definition 4.2.(4) of the receptance namely

$$(9) \qquad \underline{x}(t) = \alpha(if).\quad P_0 e^{i2\pi ft} ,$$

we find the important relation

$$\alpha(if) = \int_{-\infty}^{+\infty} W(\tau)\, e^{-i2\pi f\tau}\, d\tau \,, \tag{10}$$

i.e. the receptance $\alpha(if)$ is the Fourier transform of the impulsive receptance $W(\tau)$. The Fourier theorem in 3.3 then provides the inverse relation

$$W(\tau) = \int_{-\infty}^{+\infty} \alpha(if)\, e^{i2\pi f\tau}\, df \,. \tag{11}$$

The existence of transformations (10) and (11) is what is meant by saying $W(\tau)$ and $\alpha(if)$ are a Fourier transform pair.

5.5. Examples of relations 5.4 (10,11)

(1) <u>One degree of freedom</u> $\ddot{x} + 2\zeta\omega_0\dot{x} + \omega_0^2 x = P(t)$.

The impulsive receptance for $P(t) = \delta(t)$, i.e. for unit mass m, is

$$W(t) = \frac{1}{q}\, e^{-pt} \sin qt \,,$$

where $p = \zeta\omega_0 = \zeta 2\pi f_0$, $q = \omega_0(1-\zeta^2)^{\frac{1}{2}} = 2\pi f_0(1-\zeta^2)^{\frac{1}{2}}$.

According to 5.4(10) we have

$$\alpha(if) = \int_{0}^{\infty} \frac{1}{q}\, e^{-pt} \sin qt\, e^{-i2\pi ft}\, dt \,.$$

Denoting the real part of the receptance by $R\alpha$ and the imaginary part by $I\alpha$ we have

$$R\alpha = \int_0^\infty \frac{1}{q} e^{-pt} \sin qt \cos 2\pi ft \, dt$$

$$= \frac{1}{2q} \int_0^\infty e^{-pt} \sin(q + 2\pi f)t \, dt + \frac{1}{2q} \int_0^\infty e^{-pt} \sin(q - 2\pi f)t \, dt .$$

Similarly

$$I\alpha = \frac{1}{2q} \int_0^\infty e^{-pt} \cos(q - 2\pi f)t \, dt - \frac{1}{2q} \int_0^\infty e^{-pt} \cos(q + 2\pi ft)t \, dt .$$

These four integrals can be evaluated using the identities

$$\int_0^\infty e^{-pt} \sin rt \, dt = \frac{r}{p^2 + r^2} , \qquad \int_0^\infty e^{-pt} \cos rt \, dt = \frac{p}{p^2 + r^2} .$$

After quite some algebraic rearrangement it is finally found that

$$\alpha(if) = R\alpha + iI\alpha = \frac{1}{(p^2 + q^2 - 4\pi^2 f^2) + i2p(2\pi f)}$$

$$= \frac{1}{4\pi^2 f_0^2 - 4\pi^2 f^2 + i2\zeta(2\pi f_0)(2\pi f)}$$

$$= \frac{1}{\omega_0^2 - \omega^2 + i2\zeta\omega\omega_0}$$

as in 4.2(5) with $m = 1$. The above detailed derivation verifies
that the two well-known expressions for the receptances of a
one degree of freedom system are in fact a Fourier transform

pair.

(2) Integrator with finite memory

In this case the impulsive receptance is

$$W(t) = \begin{cases} 1 & \text{for } 0 < t < T \\ 0 & \text{elsewhere} \end{cases} .$$

According to 5.4(10) we have

$$\alpha(if) = \int_0^1 1\, e^{-i2\pi ft}\, dt$$

$$= \frac{\sin 2\pi fT}{2\pi f} - i\frac{(1 - \cos 2\pi fT)}{2\pi f} .$$

For the perfect integrator we know that $|\alpha| = \dfrac{1}{2\pi f}$. However, such an integrator does not satisfy the stability condition that $\int_0^\infty |W(t)|\, dt$ should be bounded, so we do not expect relation 5.4(10) to hold as $T \to \infty$.

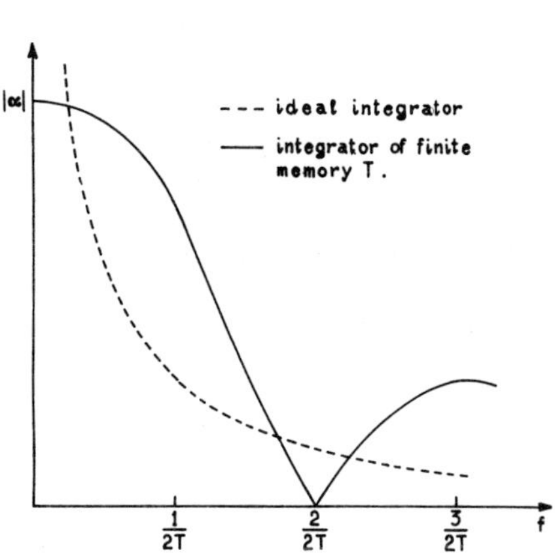

(3) Low-pass filter

The usual description of a low-pass filter is that $|\alpha(if)| = 1$ for $|f| < f_1$ and $|\alpha(if)| = 0$ for $|f| > f_1$. Here we wish to show that the corresponding impulsive receptance depends very

much on how the phase of $\boldsymbol{\alpha}$ varies in the pass band. (This fact
is often overlooked when choosing a filter). Suppose $\boldsymbol{\alpha}(if) =$
$= e^{-iaf}$ in the pass band. .

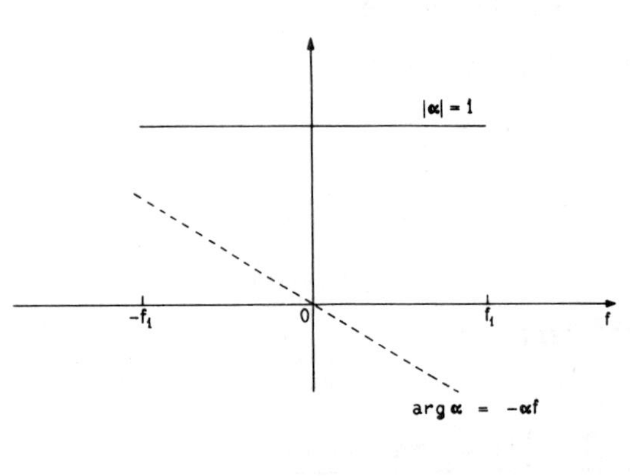

Let us determine the cor-
responding $W(t)$ using
5.4(11):

$$W(t) = \int_{-f_1}^{+f_1} e^{-iaf} e^{i2\pi ft}\, df$$

$$= \int_{-f_1}^{f_1} \cos(-a + 2\pi t) f\, df$$

$$= 2\frac{\sin(-a + 2\pi t)f_1}{(-a + 2\pi t)} = 2\frac{\sin(-2\pi t + a)f_1}{(-2\pi t + a)} \; .$$

To see what a radical effect the phase characteristic has on $W(t)$
consider the two rather extreme cases.

<u>Case 1.</u> $a = 0$

$$W(t) = 2\frac{\sin 2\pi f_1 t}{2\pi t} \; .$$

Case 2. $a = \pi/2f_1$,

$W(t)$ now shifted

distance $1/4f_1$ to

right.

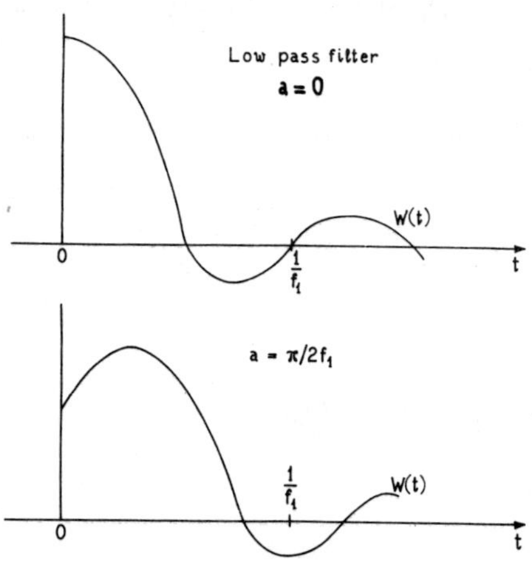

(4) <u>Spring–damper</u>

<u>system</u>

$$\alpha(if) = \frac{1}{k + i\omega c}$$

$$W(t) = \int_{-\infty}^{+\infty} \frac{1}{k + i\omega f c}\, e^{i2\pi ft}\, df$$

$$= \frac{1}{c}\, e^{-kt/c} \ .$$

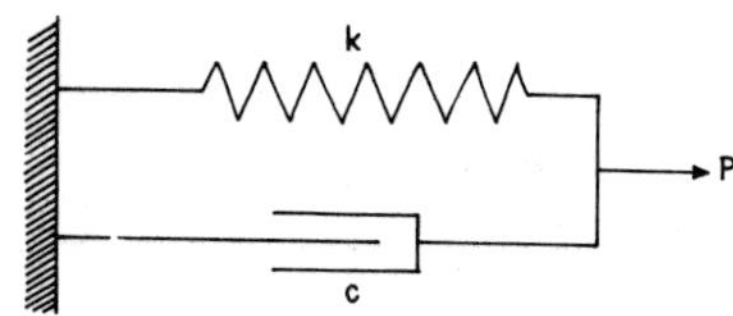

The behaviour of

α for case 2 in the preceding

example is an approximation to

that of the spring–damper sys-

tem. However, in one case $W(t)$

is oscillatory whereas in the

other it is periodic, i.e.

good approximation to α

does not imply good pointwise

approximation to $W(t)$, and con

versely.

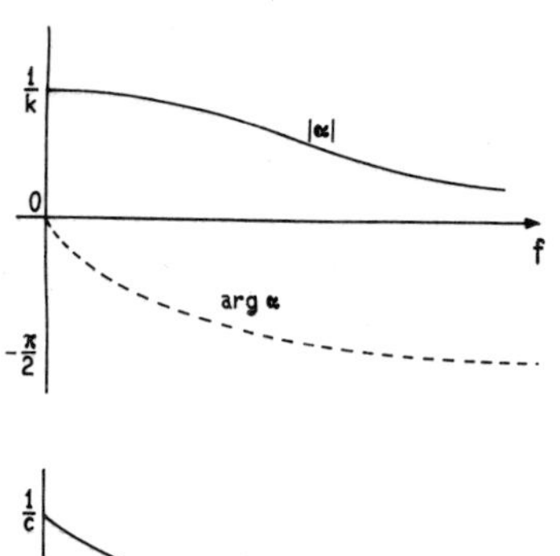

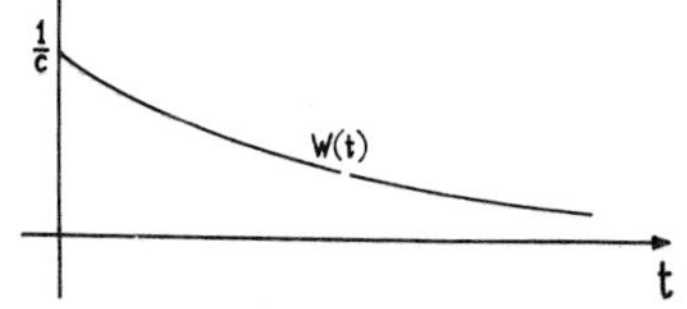

$$\text{Chapter 6}$$

DESCRIPTION III : AUTOCORRELATION FUNCTION

6.1. Autocorrelation and spectral density

We now show that the spectral density $S(f)$ is the Fourier transform of $2R(\tau)$: that is

$$S(f) \;=\; \int_{-\infty}^{\infty} 2R(\tau)e^{-i2\pi f\tau}\,d\tau\;,\;\text{with}\;\;R(\tau)\;=\;\lim_{T\to\infty}\frac{1}{T}\int_{-\frac{1}{2}T}^{\frac{1}{2}T}x(t)x(t+\tau)\,dt\,.$$

Making use again of our function $x_T(t)$ of restricted range, which is equal to $x(t)$ in the range $-\frac{1}{2}T$ to $\frac{1}{2}T$, and 0 outside,

$$\int_{-\infty}^{\infty} R_T(\tau)e^{-i2\pi f\tau}\,d\tau \;=\; \int_{-\infty}^{\infty}\left\{\frac{1}{T}\int_{-\infty}^{\infty}x_T(t)x_T(t+\tau)\,dt\right\}e^{-i2\pi f\tau}\,d\tau$$

$$=\;\frac{1}{T}\int_{-\infty}^{\infty}\int_{-\infty}^{\infty}x_T(t)x_T(t+\tau)e^{-i2\pi f(t+\tau)}e^{i2\pi ft}\,dt\,d\tau$$

$$=\;\frac{1}{T}\int_{-\infty}^{\infty}\left\{\int_{-\infty}^{\infty}x_T(t)e^{i2\pi ft}\,dt\right\}x_T(s)e^{-i2\pi fs}\,ds$$

$$\text{with}\;\;s=t+\tau$$

$$=\;\frac{1}{T}A_T(-if)A_T(if)$$

$$\text{(a)}\qquad\qquad\qquad =\;\frac{1}{T}\left|A_T(if)\right|^2$$

$$\therefore \int_{-\infty}^{\infty} R(\tau)e^{-i2\pi f\tau}\,d\tau \;=\; \lim_{T\to\infty}\left[\frac{1}{T}|A_T(if)|^2\right] \;=\; \frac{1}{2}S(f)\,. \qquad (b)$$

We have also the inverse relation: $R(\tau) = \int_{-\infty}^{\infty} \frac{1}{2} S(f)e^{i2\pi f\tau}\,df$.
This important relation between these two quantities was dis-
covered separately by Wiener and by Khinchin, and is often referred to by their names. This can be taken as the definition
of S if desired, and the other properties deduced from it.

We may repeat here that the three quantities $\langle x^2(t)\rangle$,
$R(\tau)$, and $S(f)$, are all defined in terms of a single realisation, of infinite duration, of a random process; but if the
process is stationary and ergodic, they are characteristic of
the process as a whole. The autocorrelation function and the
spectral density measure essentially the same property of a
random process, and we use whichever is most convenient for our
purpose at the time. Physically, the relation between these two
forms is, that the presence of high frequency components in the
random process, as shown in the spectral density, destroys the
correlation between successive values of x except at short time
intervals, as shown by the autocorrelation function decreasing
rapidly with τ .

6.2. Examples of spectral density

We illustrate the meaning of these quantities by applying our analysis first to determinate functions, and then to idealised random functions. A summary of the relations which we have obtained so far is given in an appendix to this chapter, together with some examples of Fourier transforms which we shall require, and a note on the impulse function $\delta(t)$.

6.3. Sinusoidal function

Let $x(t) = a\cos(2\pi f_0 t + \Phi)$

where f_0 is a constant frequency and Φ is the phase angle

$$\langle x^2(t)\rangle = \lim_{T\to\infty} \frac{1}{T}\int_{-\frac{1}{2}T}^{\frac{1}{2}T} a^2\cos^2(2\pi f_0 t + \Phi)dt = \frac{1}{2}a^2$$

$$R(\tau) = \lim_{T\to\infty}\frac{1}{T}\int_{-\frac{1}{2}T}^{\frac{1}{2}T} a^2\cos(2\pi f_0 t + \Phi)\cos(2\pi f_0(t+\tau)+\Phi)dt$$

$$= \frac{1}{2}a^2\cos 2\pi f_0\tau$$

$$S(f) = 2\int_{-\infty}^{\infty}\frac{1}{2}a^2\cos 2\pi f_0\tau\, e^{-i2\pi f\tau}d\tau$$

$$= \frac{a^2}{2}\int_{-\infty}^{\infty}\left[e^{-i2\pi(f-f_0)\tau} + e^{-i2\pi(f+f_0)\tau}\right]d\tau$$

$$= \frac{1}{2}a^2\left[\delta(f-f_0) + \delta(f-f_0)\right].$$

This represents isolat‐
ed peaks, of area $\frac{1}{2}a^2$,
at the points $f = f_0$ and
$f = f_0$. As a check

$$\langle x^2(t)\rangle = \int_0^\infty S(f)\,df = \frac{1}{2}a^2 .$$

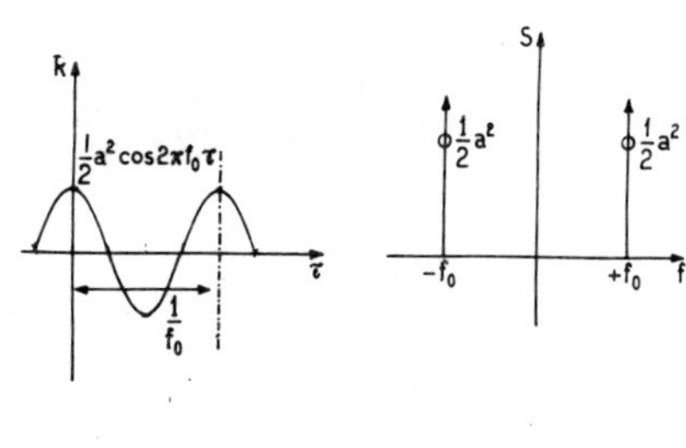

Fig. 6.1

These functions are shown in Fig. 6.1.

We note that a periodic component in $x(t)$ appears as a periodic component in $R(\tau)$ and as a pair of isolated peaks in $S(f)$. The phase angle ϕ does not appear at all.

6.4. General Periodic function

We can write down at once the corresponding re‐
sults for any function which is represented by a Fourier series,
the summation of terms in R being justified because the cross‐
correlation functions are zero.

$$x(t) \;=\; a_0 + \sum_1^\infty a_n \cos 2\pi n f_0 t + \sum_1^\infty b_n \sin 2\pi n f_0 t \;=\; \sum_{-\infty}^\infty c_n e^{i2\pi n f_0 t}$$

$$\langle x^2(t)\rangle \;=\; a_0^2 + \sum_1^\infty \frac{1}{2}(a_n^2 + b_n^2) \;=\; \sum_{-\infty}^\infty |c_n|^2$$

$$R(\tau) \;=\; a_0^2 + \sum_1^\infty \frac{1}{2}(a_n^2 + b_n^2)\cos 2\pi n f_0 \tau \;=\; \sum_{-\infty}^\infty |c_n|^2 \cos 2\pi n f_0 \tau$$

$$S(f) \;=\; 2a_0^2\,\delta(f) + \sum_1^\infty \frac{1}{2}(a_n^2 + b_n^2)\big[\delta(f - nf_0) + \delta(f + nf_0)\big]$$

$$=\; \sum_{-\infty}^{\infty}|c_n|^2\big[\delta(f - nf_0) + \delta(f + nf_0)\big]\,.$$

6.5. Square wave

As an application of the preceding result, suppose x is a square wave of amplitude A and period T .

Then
$$x(t) \;=\; \frac{4A}{\pi}\left\{\sin 2\pi\,\frac{t}{T} + \frac{1}{3}\sin 6\pi\,\frac{t}{T} + \frac{1}{5}\sin 10\pi\,\frac{t}{T} + \ldots\right\}$$

$$R(\tau) \;=\; \frac{8A^2}{\pi^2}\left\{\cos 2\pi\,\frac{\tau}{T} + \frac{1}{3^2}\cos 6\pi\,\frac{\tau}{T} + \frac{1}{5^2}\cos 10\pi\,\frac{\tau}{T} + \ldots\right\}$$

which represents a triangular wave of amplitude A^2 and period T.

$$S(f) \;=\; \frac{8A^2}{\pi^2}\left\{\delta\!\left(f - \frac{1}{T}\right) + \delta\!\left(f + \frac{1}{T}\right) + \frac{1}{3^2}\delta\!\left(f - \frac{3}{T}\right) + \frac{1}{3^2}\delta\!\left(f + \frac{3}{T}\right) + \ldots\right\}$$

which represents a series of pairs of isolated peaks of diminish‐ing area, shown in Fig. 6.2.

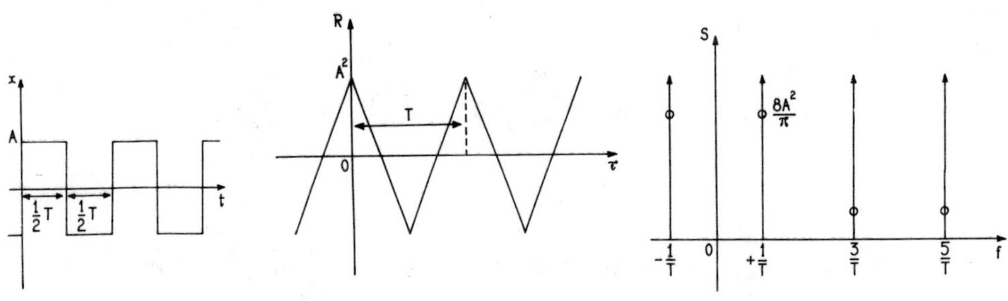

Fig. 6.2

6.6. Random pulse signal

Let $x(t)$ have the form shown in Fig. 6.3 defined

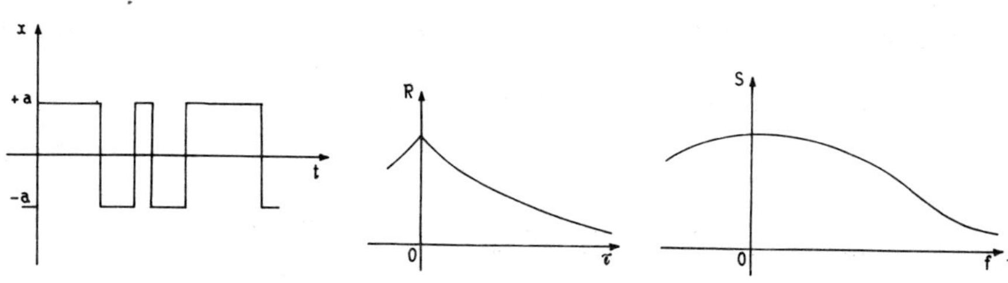

Fig. 6.3

by the following property: x is either $+a$ or $-a$, the changes of sign occurring at random times, with an average time interval T_m . The events (changes of sign) occur according to a Poisson probability distribution. We shall discuss this later, and for the moment we quote the result that the probability of the occurrence of k events in a specified time T is given by:

$$P(k) = \frac{(\lambda T)^k}{\lfloor k} e^{-\lambda T} , \qquad \text{where} \qquad \lambda = \frac{1}{T_m} .$$

The auto-correlation function is the average value of $x(t)x(t + \tau)$. This product will have the value $-a^2$ when τ is chosen so that there are an odd number 1, 3, 5, etc. of changes of sign between t and $t + \tau$, and the product will be $+a^2$ for an even number 0, 2, 4, etc. changes.

So $<x(t)x(t + \tau)> = a^2 \Pr[\text{even number}] - a^2 \Pr[\text{odd number}]$

$$= a^2 [P(0) + P(2) + \ldots] - a^2 [P(1) + P(3) + \ldots] .$$

Using the formula above this will be

$$a^2\left[e^{-\lambda\tau} + \frac{(\lambda\tau)^2}{\lfloor 2}e^{-\lambda\tau} + \ldots\right] - a^2\left[(\lambda\tau)e^{-\lambda\tau} + \frac{(\lambda\tau)^3}{\lfloor 3}e^{-\lambda\tau} + \ldots\right] = e^{-2\lambda\tau}a^2.$$

We therefore write $R(\tau) = a^2 e^{-2\lambda|\tau|}$, since R is an even function. The mean square value is of course a^2.

$$\text{Spectral density} \quad S(f) = 2\int_{-\infty}^{\infty} a^2 e^{-2\lambda|\tau|} e^{-i2\pi f\tau}\, d\tau$$

$$= 4a^2 \int_{0}^{\infty} e^{-2\lambda\tau} \cos 2\pi f\tau\, d\tau$$

$$= \frac{2a^2\lambda}{\pi^2 f^2 + \lambda^2}.$$

This pair of autocorrelation function and spectral density are of frequent occurrence in theoretical work.

6.7. Specified autocorrelation function

It is sometimes convenient to start from a specified form of $R(\tau)$. We have just shown that $R(\tau) = a^2 e^{-k|\tau|}$ leads to the form:

$$S(f) = \frac{4a^2 k}{4\pi^2 f^2 + k^2}.$$

If we take $R(\tau) = a^2 e^{-k|\tau|}\cos 2\pi f_0 t$, and use the Fourier transform of this function given in the table, we find

$$S(f) = \frac{4a^2 k\left[4\pi^2(f^2 + f_0^2) + k^2\right]}{\left[4\pi^2(f^2 + f_0^2) + k^2\right]^2 - 64\pi^4 f^2 f_0^2} \ .$$

If k is not too large, this has a maximum near $f = f_0$, as shown in Fig. 6.4, represent-

ing a predominating group of frequencies in $x(t)$ in the neighbourhood of this value.

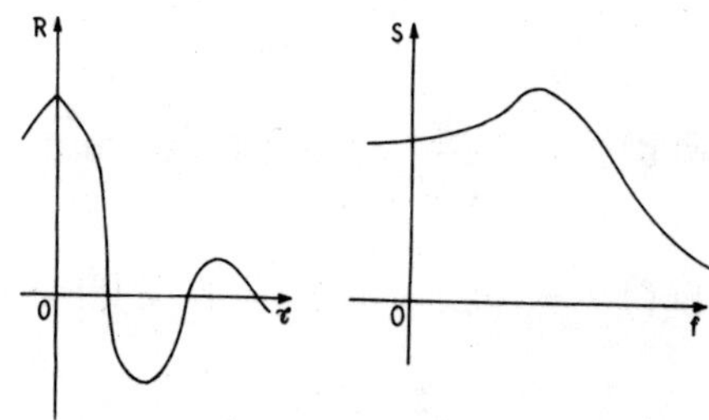

Fig. 6.4

6.8. Uniform spectral density

If we wish to start from a specified spectral density, it is natural to think of a uniform value: $S(f) = a^2$, independent of f. This would give $R(\tau) = \frac{1}{2}a^2 \delta(\tau)$, representing a peak of area $\frac{1}{2}a^2$ at $\tau = 0$, and 0 elsewhere. In fact, however, this bears no resemblance to our usual image of a random function since $x(t)$ would be dominated by indefinitely high frequencies; evidently the mean square has an infinite value. Nevertheless it is often found useful to use this form as a hypothetical input to a system when we are really interested in the output. Every system ultimately rejects the highest frequencies to which it cannot respond; and, provided we have a plausible specification for the input over the effective frequency range, it does not much matter what input we propose beyond this point. Such an input is called 'white noise'.

We can of course specify $S(f)$ as constant over a certain frequency range, and zero outside it. This may be termed bandlimited white noise. This gives the following for $R(\tau)$.

(1) $S(f) \;=\; a^2$ for $\quad |f| < f_1; \quad 0 \quad$ for $\quad |f| > f_1$

$$R(\tau) \;=\; \frac{a^2}{2\pi\tau}\sin 2\pi f_1\tau \quad \text{and} \quad <x^2(t)> \;=\; a^2 f_1$$

(2) $S(f) \;=\; a^2 \quad$ for $\quad f_2 < |f| < f_1; \quad 0 \quad$ for $\quad |f| < f_2; \quad 0 \quad$ for $|f| > f_1$

$$R(\tau) \;=\; \frac{a^2}{2\pi\tau}\left[\sin 2\pi f_1\tau - \sin 2\pi f_2\tau\right]; \quad <x^2(t)> \;=\; a^2(f_1 - f_2)$$

(3) as case (2), with a narrow frequency band of width δf_0 between f_2 and f_1 .

$$\text{Put} \quad f_1 \;=\; f_0 + \frac{1}{2}\delta f_0 \quad \text{and} \quad f_2 \;=\; f_0 - \frac{1}{2}\delta f_0$$

$$R(\tau) \;=\; \frac{a^2}{\pi\tau}\cos 2\pi f_0\tau \, \sin \pi(\delta f_0)\tau$$

$$<x^2(t)> \;=\; a^2(\delta f) .$$

APPENDIX FOR CHAPTER 6

1. Summary of relation between characteristics of random processes

Mean square value of x: $<x^2(t)>$

Autocorrelation function: $R(\tau)$

Spectral density: $S(f)$

Fourier transform: $A(if)$

$A(if)$ is defined as $\displaystyle\int_{-\infty}^{\infty} x(t)e^{-i2\pi ft}\,dt$; then $x(t) = \displaystyle\int_{-\infty}^{\infty} A(if)e^{i2\pi ft}\,df$

$<x^2(t)>$ is defined as $\displaystyle\lim_{T\to\infty}\frac{1}{T}\int_{-\frac{1}{2}T}^{\frac{1}{2}T} x^2(t)\,dt$

$R(\tau)$ is defined as $\displaystyle\lim_{T\to\infty}\frac{1}{T}\int_{-\frac{1}{2}T}^{\frac{1}{2}T} x(t)x(t+\tau)\,dt$

$S(f)$ is defined as $\displaystyle\lim_{T\to\infty}\frac{2}{T}\left|A_T(if)\right|^2$, with $A_T(if) = \displaystyle\int_{-\infty}^{\infty} x_T(t)e^{-i2\pi ft}\,dt$

$x_T(t) = x(t)$ if $-\frac{1}{2}T < t < \frac{1}{2}T$, and $x_T(t) = 0$ if $|t| > \frac{1}{2}T$

$$<x^2(t)> = R(0) = \int_{0}^{\infty} S(f)\,df$$

$$S(f) = 2\int_{-\infty}^{\infty} R(\tau)e^{-i2\pi f\tau}\,d\tau$$

$$R(\tau) = \frac{1}{2}\int_{-\infty}^{\infty} S(f)e^{i2\pi f\tau}\,df\ .$$

2. Unit impulse function

The unit impulse function (Dirac delta function), written $\delta(t)$, is represented by a graph which is zero everywhere except in the neighbourhood of the point $t = 0$ where it has a peak having an area of unity. The duration, amplitude and form of the peak are not prescribed, except that the duration is indefinitely small. It obviously corresponds to the concept of an impulse in dynamics. The notation $\delta(t - t_0)$ represents a similar peak in the neighbourhood of the point $t = t_0$. $\delta(t)$ is not a 'function' in the usual mathematical sense; but within the limits of the definition it can be freely used in equations, and even provided with a derivative and an integral. A formal justification of the procedure is given in Kaplan (Ref. 3.2), who calls these things 'generalized functions'.

We have the general result that $\int_{-\infty}^{\infty} F(t) \cdot \delta(t - t_0) \, dt = F(t_0)$ for any function $F(t)$.

From this we can deduce the Fourier transforms

$$\int_{-\infty}^{\infty} \delta(t) e^{-i2\pi ft} \, dt = 1 \qquad \text{(as a function of } f \text{) and the inverse}$$

$$\int_{-\infty}^{\infty} 1 \, e^{i2\pi ft} \, df = \delta(t) \, ,$$

with the corresponding pair

$$\int_{-\infty}^{\infty} \delta(f) e^{-i2\pi ft} \, df = 1 \qquad \text{(as a function of } t \text{), and}$$

$$\int_{-\infty}^{\infty} 1 \, e^{i2\pi ft} \, dt = \delta(f) \, .$$

3. Table of Fourier transforms (double infinite integral)

Definition $\quad G(f) = \int_{-\infty}^{\infty} g(t)\,e^{-i2\pi ft}\,dt \quad$ and $\quad g(t) = \int_{-\infty}^{\infty} G(f)\,e^{i2\pi ft}\,dt$

$g(t)$	$G(f)$
1	$\delta(f)$
$\delta(t)$	1
$e^{-k\lvert t\rvert} \quad$ with $\quad k>0$	$\dfrac{2k}{4\pi^2 f^2 + k^2}$
$e^{-k\lvert t\rvert}\cos 2\pi f_0 t \quad$ with $\quad k>0$	$\dfrac{2k\left[4\pi^2(f^2 + f_0^2) + k^2\right]}{\left[4\pi^2(f^2 + f_0^2)+k^2\right]^2 - 64\pi^2 f^2 f_0^2}$
$e^{-a^2 t^2}$	$\dfrac{\sqrt{\pi}}{a}\,e^{-\frac{\pi^2 f^2}{a^2}}$
$\dfrac{1}{a^2 + t^2} \quad$ with $\quad a>0$	$\dfrac{\pi}{a}\,e^{-a2\pi\lvert f\rvert}$
e^{iat}	$\delta(f - a)$
$\cos 2\pi f_0 t$	$\dfrac{1}{2}\delta(f - f_0) + \dfrac{1}{2}\delta(f + f_0)$
$\sin 2\pi f_0 t$	$\dfrac{i}{2}\delta(f + f_0) - \dfrac{i}{2}\delta(f - f_0)\,.$

Chapter 7

RANDOM VIBRATION RESPONSE I : GENERAL RELATIONSHIPS

7.1. Introduction

We can now bring together the results of the preceding chapters to study the effect of a random input to a system, of which the output is related to the input through a linear differential equation with constant coefficients. We deal first with the situation in which there is a single random input, for example a force applied at a single point in a constant direction, although we may be interested in more than one output. Because of the complications of the superposition of random processes, referred to in § 2.11, we defer the case of multiple inputs till later (chapter 10).

We shall find that the auto-correlation functions of input and output are linked by the impulsive receptance of the system, and the spectral densities are linked by the receptance.

7.2. Response of linear system to random input

Let the input $P(t)$ be a stationary, ergodic random function, with an auto-correlation function

$$R_P(\tau) \;=\; <P(t)\,P(t + \tau)> \, .$$

Assuming that the output is a similar random function $x(t)$, we have

$$R_x(\tau) \;=\; <x(t)x(t + \tau)> \; .$$

The property of the system is defined by the impulsive receptance $W(t)$ introduced in a preceding chapter Then the output x for any input P is $x(t) = \int_0^\infty W(\tau)P(t - \tau)d\tau$.

$$R_x(\tau) \;=\; < \int_0^\infty W(\tau_1)P(t - \tau_1)d\tau_1 \int_0^\infty W(\tau_2)P(t - \tau_2 + \tau)d\tau_2> \; .$$

The dummy variables are written τ_1 and τ_2 here to maintain the distinction between them.

The time-average implies an integration with respect to t , so if we change the order of integration we get

$$R_x(\tau) \;=\; \int_0^\infty W(\tau_1)\left[\int_0^\infty W(\tau_2)<P(t - \tau_1)P(t - \tau_2 + \tau)>d\tau_2\right]d\tau_1$$

$$=\; \int_0^\infty W(\tau_1)\left[\int_0^\infty W(\tau_2)R_P(\tau_1 - \tau_2 + \tau)d\tau_2\right]d\tau_1$$

giving R_x in terms of R_P and W . The fact that R_x depends only on τ and not on t confirms that the output is a stationary random process.

We now use this result to link the spectral densities.

$$S_P(f) = 2\int_{-\infty}^{\infty} R_P(\tau)e^{-i2\pi f\tau}\,d\tau \quad \text{and} \quad S_x(f) = 2\int_{-\infty}^{\infty} R_x(\tau)e^{-i2\pi f\tau}\,d\tau$$

$$S_x(f) = 2\int_{-\infty}^{\infty}\left[\int_0^{\infty} W(\tau_1)\int_0^{\infty} W(\tau_2)R_P(\tau_1 - \tau_2 + \tau)\,d\tau_2\,d\tau_1\right]e^{-i2\pi f\tau}\,d\tau$$

$$= 2\int_0^{\infty} W(\tau_1)e^{i2\pi f\tau_1}\,d\tau_1 \cdot \int_0^{\infty} W(\tau_2)e^{-i2\pi f\tau_2}\,d\tau_2\,.$$

$$\cdot \int_{-\infty}^{\infty} R_P(\tau_1 - \tau_2 + \tau)e^{-i2\pi f(\tau_1 - \tau_2 + \tau)}\,d(\tau_1 - \tau_2 + \tau)\,.$$

We know that $\alpha(if) = \int_0^{\infty} W(\tau)e^{-i2\pi f\tau}\,d\tau$

so $S_x(f) = 2\,\alpha(-if)\alpha(if)\frac{1}{2}S_P(f)$.

Also $\quad \alpha(-if) = \alpha^{*}(if)$, the conjugate of $\alpha(if)$

$$\therefore S_x(f) = \left|\alpha(if)\right|^2 S_P(f)\,.$$

In this expression, we use only $\left|\alpha(if)\right|$ as a property of the system, that is the magnitude ratio of output to input for sinusoidal input. The phase relation implied in $\alpha(if)$ as a complex number is not required.

7.3. Mean and variance

We can also obtain expressions for the relation between the mean value and variance of the output and the input.

$$\langle P(t)\rangle = \lim_{T\to\infty}\frac{1}{T}\int_{-\frac{1}{2}T}^{\frac{1}{2}T} P(t)\,dt, \text{ with a similar formula for } \langle x(t)\rangle\,.$$

Then
$$<x(t)> = \lim_{T \to \infty} \frac{1}{T} \int_{-\frac{1}{2}T}^{\frac{1}{2}T} \left[\int_{0}^{\infty} W(\tau)P(t-\tau)d\tau \right] dt$$

$$= \int_{0}^{\infty} W(\tau)<P(t-\tau)>d\tau$$

$$= <P(t)> \int_{0}^{\infty} W(\tau)d\tau$$

since $<P(t)>$ is a number, independent of t and τ.

But we have: $\alpha(if) = \int_{0}^{\infty} W(\tau)e^{-i2\pi f\tau}d\tau$

so that $\int_{0}^{\infty} W(\tau)d\tau = \alpha(0)$

$$\therefore <x(t)> = \alpha(0)<P(t)> \, ,$$

$\alpha(0)$ is a real number.

Also
$$<x^2(t)> = \int_{0}^{\infty} S_x(f)df$$

$$= \int_{0}^{\infty} |\alpha(if)|^2 S_P(f)df \, .$$

If the input process has a Gaussian probability density, as described in § 2.5, so has the output; and these two relations link the parameters in the density functions for output and input, so that we have a complete knowledge of the output density function. If the input is not Gaussian, these equations are still valid, but we do not know the nature of the output density function, apart from the values of its mean and variance.

7.4. Random input to first order mechanical system

This is represented by a system with spring and viscous damping, but with negligible mass, so that the equation takes the form

$$c\dot{x} + kx = P(t) \quad , \quad \text{for one degree of freedom.}$$

The impulsive receptance for this system is

$$W(t) = \frac{e^{-\mu t}}{c} \quad \text{with} \quad \mu = k/c .$$

The receptance is
$$\alpha(if) = \frac{1}{k + i2\pi fc} .$$

Using the auto-correlation function

$$R_x(\tau) = \frac{1}{c^2}\int_0^\infty e^{-\mu\tau_1}\left[\int_0^\infty e^{-\mu\tau_2} R_P(\tau_1 - \tau_2 + \tau)d\tau_2\right]d\tau_1 .$$

For white noise input $R_P(\tau) = \frac{1}{2}a^2\cdot\delta(\tau)$.

$$\therefore R_x(\tau) = \frac{a^2}{2c^2}\int_0^\infty e^{-\mu\tau_1}\int_0^\infty e^{-\mu\tau_2}\,\delta(\tau_1 - \tau_2 + \tau)d\tau_2\,d\tau_1$$

$$= \frac{a^2}{2c^2}\int_0^\infty e^{-\mu\tau_1}e^{-\mu(\tau_1+\tau)}d\tau_1$$

(a)

$$= \frac{a^2}{4ck}e^{-\mu|\tau|} .$$

$$<x^2(t)> \;=\; \frac{a^2}{4ck} \qquad\qquad (b)$$

so the mean square of the output is finite, although that of the input would be infinite. This arises from the attenuation of the higher frequencies, which is shown more clearly by considering the spectral density.

From the above, $\left|\alpha(if)\right|^2 = \dfrac{1}{k^2 + 4\pi^2 f^2 c^2}$, so $S_x(f) = \dfrac{S_P(f)}{k^2 + 4\pi^2 f^2 c^2}$

For white noise input $S_P(f) = a^2$ and $S_x(f) = \dfrac{a^2}{k^2 + 4\pi^2 f^2 c^2}$.

These will be recognised as the associated pair of R and S previously discussed (§ 6.6).

For white noise input, $<x^2(t)> = a^2 \displaystyle\int_0^\infty \left|\alpha(if)\right|^2 \cdot df$,

which gives $\dfrac{a^2}{4\,ck}$ as above.

Integrals of this type are of frequent occurrence and are tedious to work out for more complex systems; a table of some standard types is given in Crandall and Mark, page 72 (Ref. 1.1).

If we suppose the input to be of the type

$$R_P(\tau) = a^2 e^{-b|\tau|} \;,\quad S_P(f) = \frac{4a^2 b}{b^2 + 4\pi^2 f^2} \;,\quad \text{applied to same system,}$$

we obtain $R_x(\tau) = \dfrac{a^2}{k^2 - b^2 c^2}\left[e^{-b|\tau|} - \left(\dfrac{bc}{k}\right)e^{-\mu|\tau|}\right]$

$$\text{and} \quad S_x(f) = \frac{4a^2 b}{(b^2 + 4\pi^2 f^2)(k^2 + 4\pi^2 f^2 c^2)}$$

$$\text{These give} \quad <x^2(t)> = \frac{a^2}{k(k + bc)} \ .$$

[Note that the integrals for R_x have to be done in sections since R_P is represented by different algebraic forms for positive and for negative values of the argument] .

7.5. Random input to second order mechanical system

In the previous example, the input is attenuated at all frequencies in passing through the system; but a system of second order has a natural frequency of its own, and this will be reflected in the output unless the damping is too large. The situation as it affects the spectral density is shown in Fig. 7.1.

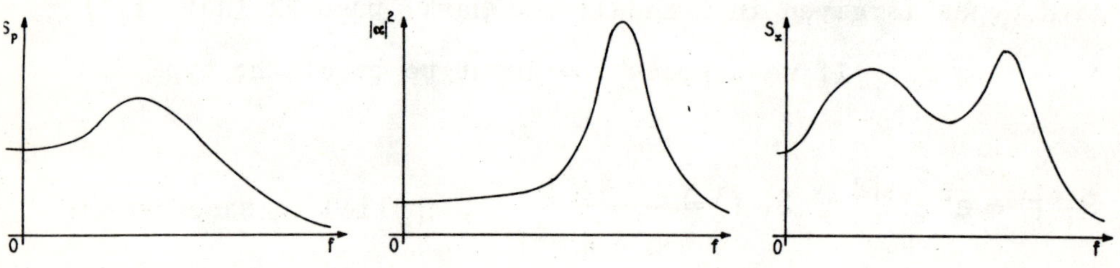

Fig. 7.1

Taking a conventional system consisting of a mass controlled by a spring and viscous damper, the equation of motion will be

$$m\ddot{x} + c\dot{x} + kx = P(t).$$

This gives the impulsive receptance

$$W(t) = \frac{e^{-2\pi\zeta f_0 t}\cdot\sin 2\pi f_d t}{2\pi m f_d}$$

if the damping is small $(\zeta < 1)$ so that the response to an impulse is oscillatory.

Here the natural frequency is $f_0 = \dfrac{\sqrt{k}}{2\pi\sqrt{m}}$,

damping factor $\zeta = \dfrac{c}{4\pi m f_0}$

frequency of damped vibration $f_d = f_0\sqrt{1-\zeta^2}$.

For a white noise input with $R_P(\tau) = \frac{1}{2}a^2\cdot\delta(t)$, this gives

$$R_x(\tau) = \frac{a^2 e^{-2\pi\zeta f_0|\tau|}\cdot\cos(2\pi f_d|\tau|-\Phi)}{4ck\sqrt{1-\zeta^2}} \quad\text{with}\quad \cos\Phi = \sqrt{1-\zeta^2}.$$

This gives $<x^2(t)> = R_x(0) = a^2/4ck$.

The damped frequency f_d of the system shows up in the output auto-correlation function.

The receptance for this system is :

$$\alpha(if) = \frac{1}{4\pi^2 m\left[(f_0^2 - f^2) + i2\zeta f_0 f\right]} \,,$$

so that if $S_p(f) = a^2$,

$$S_x(f) = \frac{a^2}{16\pi^4 m^2\left[(f_0^2 - f^2)^2 + 4\zeta^2 f^2 f_0^2\right]} \,.$$

We can verify that $\langle x^2(t)\rangle = a^2\int_0^\infty |\alpha(if)|^2 \cdot df = a^2/4ck$ as above

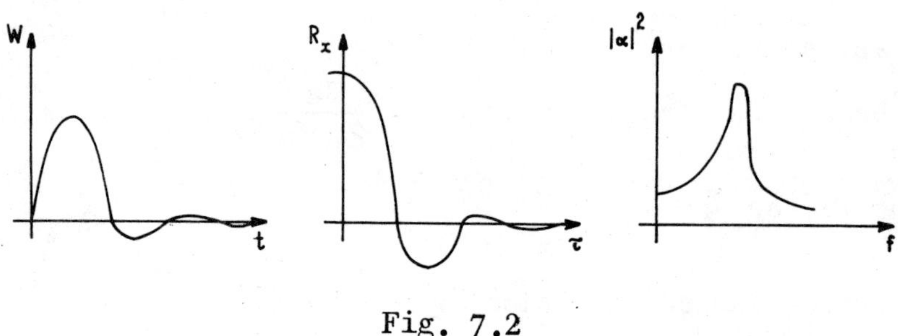

Fig. 7.2

These results are shown in Fig. 7.2. The spectral density has
a peak at frequencies in the neighbourhood of $f = f_0$, the amplitude of the peak depending on the value of ζ . The ratio
of the value of $S(f_0)$ to $S(0)$ is $1/4\zeta^2 = Q^2$ (see § 4.3). For a
system with small damping the peak will be sharp, and the output will be a random motion with the frequencies in this neighbourhood strongly predominating.

Chapter 8   VIBRATION THEORY III : NORMAL MODES

8.1. Introduction

In this chapter we will be concerned with the modal analysis of the response of lightly damped linear systems To begin with we wish to consider systems of such complexity that they cannot be regarded as a collection of concentrated masses, beams, shells, etc. whose partial differential equations can be formulated on the basis of geometric dimensions and material properties. Frequently it is found that the configuration of a machine – even as complicated as an aircraft – can be described adequately in terms of a certain set of n displacements or rotations $q_j(j=1,2,\ldots,n)$ at judiciously chosen points P_j . It is to be noted that in assuming that the q_j describe the configuration independent of time we are to a certain extent excluding systems with time varying parameters. For the sake of simplicity it will be assumed that no forced displacements are imposed on the system. Thus, in the parlance of higher dynamics, we will be considering scleronomic as opposed to rheonomic systems. After deciding on the n significant generalized coordinates q_j , the next step would be to conduct vibration tests on the given machine by applying sinusoidal forces at the points P_j . After considerable effort one will be in a position to set up a system of second-order differential equations governing the system. (The necessary size of n may have been increased or decreased during the tests). Since the usual test procedures stem from the theory of modal response based on the governing equations, we will follow the usual artificial lines of argument by starting with these governing equations

i.e. (1)
$$A\ddot{q} + B\dot{q} + Cq = Q$$
$$a_{jk}\ddot{q}_k + b_{jk}\dot{q}_k + c_{jk}q_k = Q_k.$$

The Lagrangian method of setting up equations (1) yields a mass matrix A and a stiffness matrix C that are symmetric. In addition it is clear that A must be positive definite in all real cases, since the kinetic energy of a moving physical system can never be zero or negative.

Here as in Chapter 4 we will use the complex notation to represent sinusoidally varying quantities. If the n forces in Q are all at frequency ω, we write $Q_j = (f_j + ig_j) \cdot \exp(i\omega t)$. In view of the linearity, the response will be $q_j = (u_j + iv_j)\exp(i\omega t)$, i.e., $q = (u + iv)\exp(i\omega t)$. Substitution of these expressions into (1) and cancellation of the factor $\exp(i\omega t)$ yields

(2)
$$(C + i\omega B - \omega^2 A)(u + iv) = f + ig ,$$

which is equivalent to the two real equations

$$(\underset{\sim}{C} - \omega^2\underset{\sim}{A})\underset{\sim}{u} - \omega\underset{\sim}{B}\underset{\sim}{v} = \underset{\sim}{f}\,,$$
$$\omega\underset{\sim}{B}\underset{\sim}{u} + (\underset{\sim}{C} - \omega^2\underset{\sim}{A})\underset{\sim}{v} = \underset{\sim}{g}\,.$$

(3)

Here it is to be noted in passing that, being the ratio of the complex response to the complex excitation, the receptance of the present system is given by $\underset{\sim}{\alpha}(i\omega) = (\underset{\sim}{C} + i\omega\underset{\sim}{B} - \omega^2\underset{\sim}{A})^{-1}$, which is similar in form to the corresponding expression 4.2(5) for the one freedom case.

If all the forces are in phase we can set $\underset{\sim}{g} = 0$ with $\underset{\sim}{f}$ describing the distribution of amplitudes of the excitation. The system is said to be vibrating in a <u>normal mode</u> when the frequency ω and $\underset{\sim}{f}$ are such that all the displacements have a common phase differing from the excitation by 90°. (Alternatively, the forces and velocities are in phase). Equations (3) then reduce to

$$-\omega\underset{\sim}{B}\underset{\sim}{v} = \underset{\sim}{f}\,,$$

(4)

$$(\underset{\sim}{C} - \omega^2\underset{\sim}{A})\underset{\sim}{v} = \underset{\sim}{0}\,.$$

(5)

In order that (5) have non-zero roots $\underset{\sim}{v}_k \neq 0$ it is necessary and sufficient that $\det(\underset{\sim}{C} - \omega_k^2\underset{\sim}{A}) = 0$, which is the characteristic equation for the associated conservative system $\underset{\sim}{A}\ddot{\underset{\sim}{q}} + \underset{\sim}{C}\underset{\sim}{q} = 0$. The vector $\underset{\sim}{v}_k$ is called the k-th modal shape and is obviously independent of the structure of the damping matrix $\underset{\sim}{B}$. The fact that $\underset{\sim}{A}$ and $\underset{\sim}{C}$ are non-negative and symmetric guarantees that all the roots ω_k^2 are non-negative. The zero roots correspond to rigid-body modes and are usually treated separately. Since $\underset{\sim}{A}$, $\underset{\sim}{C}$ are symmetric and one at least is positive definite, it can be proved that n linearly independent modal shapes $\underset{\sim}{v}_k$ are defined by (5) even if there are less than n distinct modal frequencies ω_k , i.e. it is possible to have more than one modal shape corresponding to one modal frequency. The probability of such degenerate cases is fairly low. Being linearly independent, the important thing about the modal shapes is that any n -dimensional vector $\underset{\sim}{v}$ can be represented as a linear combination of the n modal shapes i.e.

$$(6) \qquad \underset{\sim}{v} = a_1 \underset{\sim}{v}_1 + a_2 \underset{\sim}{v}_2 + \ldots + a_n \underset{\sim}{v}_n \ .$$

8.2. Orthogonality

Two vectors $\underset{\sim}{r}_1 = (x_1, y_1, z_1)$ and $\underset{\sim}{r}_2 = (x_2, y_2, z_2)$ are perpendicular or orthogonal when their scalar product $x_1 x_2 + y_1 y_2 + z_1 z_2$ vanishes. If the above vectors are treated as column vectors, the scalar product can be written as $\underset{\sim}{r}_1^T \underset{\sim}{r}_2$ or $\underset{\sim}{r}_2^T \underset{\sim}{r}_1$, where superscript T indicates transposition to a row vector. A weighted scalar product can be defined by $a x_1 x_2 + b y_1 y_2 + c z_1 z_2$ which in matrix notation is written

$$[x_1, y_1, z_1]\begin{bmatrix} a & 0 & 0 \\ 0 & b & 0 \\ 0 & 0 & c \end{bmatrix}\begin{bmatrix} x_2 \\ y_2 \\ z_2 \end{bmatrix} = \underset{\sim}{r}_1^T \underset{\sim}{D}\, \underset{\sim}{r}_2 .$$

Having gone so far, the last step in generalizing the idea of orthogonality is to admit a symmetric non-diagonal matrix $\underset{\sim}{D}$ for calculating the scalar product

$$\underset{\sim}{r}_1^T \underset{\sim}{D}\,\underset{\sim}{r}_2 = \underset{\sim}{r}_2^T \underset{\sim}{D}\,\underset{\sim}{r}_1 = d_{11}x_1 x_2 + d_{12}(x_1 y_2 + x_2 y_1) + \dots .$$

Thus in two dimensions the "length" of a vector (x, y) becomes $d_{11}x^2 + 2d_{12}xy + d_{22}y_2$. This is simply equivalent to using ellipses instead of circles to measure the distance from a point. Because D defines how we measure distances, it is called the <u>metric</u>.

After these preliminary remarks, let us show that the normal modes are orthogonal in this

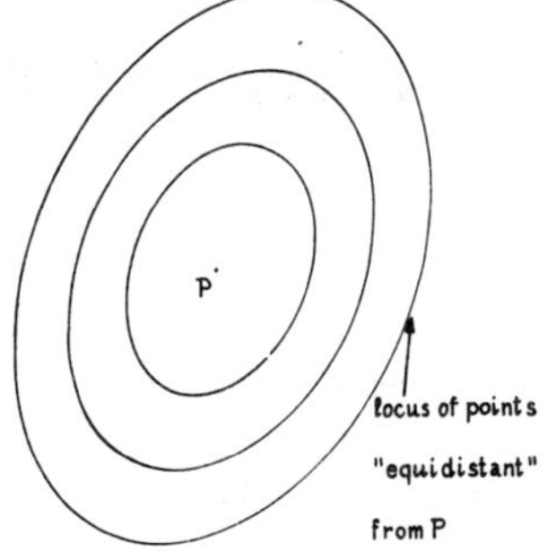

generalized sense, the mass and stiffness matrices playing the roles of metrics.

For two distinct modal frequencies ω_1 and ω_2 8.1(5) reads

$$\underset{\sim}{C}\underset{\sim}{v}_1 - \omega_1^2 \underset{\sim}{A}\underset{\sim}{v}_1 = 0, \quad \underset{\sim}{C}\underset{\sim}{v}_2 - \omega_2^2 \underset{\sim}{A}\underset{\sim}{v} = 0 . \qquad (7)$$

Premultiplying the first equation by $\underset{\sim}{v}_1^T$, the second by $\underset{\sim}{v}_2^T$

and subtracting the results we obtain

$$\omega_1^2 \, v_2^T A \, v_1 - \omega_2^2 \, v_1^T A \, v_2 \;=\; 0 \, .$$

Hence, due to the symmetry of A and since $\omega_1^2 - \omega_2^2 \neq 0$, for the modal vectors we have the orthogonality relations

$$(8) \qquad\qquad v_1^T A \, v_2 \;=\; 0 \, .$$

Moreover, the positive definiteness of A means that

$$(9) \qquad v_1^T A \, v_1 \;=\; m_1 > 0 \quad \text{and} \quad v_2^T A \, v_2 \;=\; m_2 > 0 \, .$$

Now from (7) we have the additional orthogonality relations

$$(10) \qquad v_i^T C \, v_j \;=\; \omega_j^2 \, v_i^T A \, v_j \;=\; \begin{cases} 0 & \text{when} \quad i \neq j \\[2mm] m_j \omega_j^2 & \text{when} \quad i = j \, . \end{cases}$$

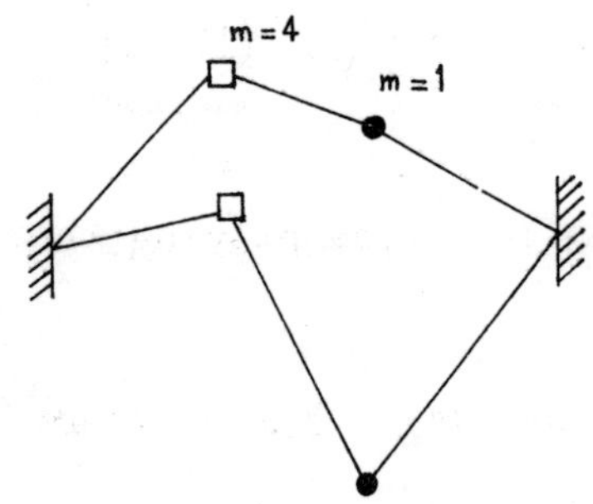

To explain the utility of the two orthogonality relations (8), (10) consider two unequal masses on an elastic cable. With a suitable choice of units we have $A = \begin{bmatrix} 4 & 0 \\ 0 & 1 \end{bmatrix}$, $C = \begin{bmatrix} 2 & -1 \\ -1 & 2 \end{bmatrix}$.

The modal vectors v_1, v_2 can be found from the orthogonality relations (8) and (10) above:

$$v_1^T \;=\; (1, \sqrt{13} - 3)$$

$$\underset{\sim}{v}_2^{\mathsf{T}} = \left(\frac{1}{2}, \frac{-\sqrt{13}-3}{2} \right)$$

The orthogonality relations yield the modal shapes without requiring the calculation of the corresponding modal frequencies.

In the case of proportional damping $\underset{\sim}{B}$ is a lin-
ear combination of $\underset{\sim}{A}$ and/or $\underset{\sim}{C}$, so in this case we have

$$\underset{\sim}{v}_j^{\mathsf{T}} \underset{\sim}{B} \underset{\sim}{v}_k = \begin{cases} 0 & \text{when} \quad j \neq k \\ 2\zeta_k m_k \omega_k & \text{when} \quad j = k . \end{cases} \tag{11}$$

Let us now see how the response $\underset{\sim}{q}$ in system (1)
can be expressed in terms of the normal modes. For the sake of
deriving a modal expression for the receptance we let the ex-
citation be $\underset{\sim}{Q} = \underset{\sim}{p} \exp(i\omega t)$. Substitution of $\underset{\sim}{q} = (a_1 \underset{\sim}{v}_1 + \dots + a_n \underset{\sim}{v}_n)$
$\exp(i\omega t)$ yields

$$\sum_{j=1}^{n} a_j \left\{ (\underset{\sim}{C} - \omega^2 \underset{\sim}{A}) + i\omega \underset{\sim}{B} \right\} \underset{\sim}{v}_j = \underset{\sim}{p} . \tag{12}$$

In view of the relations (8), (9), (10) and (11) premultiplica-
tion of both sides of (8.12) by $\underset{\sim}{v}_k^{\mathsf{T}}$ yields

$$a_k = \underset{\sim}{v}_k^{\mathsf{T}} \underset{\sim}{p} \frac{1}{m_k \left\{ (\omega_k^2 - \omega^2) + 2i\zeta_k \omega_k \omega \right\}}$$

$$\tag{13}$$

$$\underset{\sim}{q} = \sum_{j=1}^{n} \underset{\sim}{v}_j^{\mathsf{T}} \underset{\sim}{p} \frac{1}{m_j \left\{ (\omega_j^2 - \omega^2) + 2i\zeta_j \omega_j \omega \right\}} \underset{\sim}{v}_j \exp(i\omega t) .$$

From this expression it is clear that any mode $\underset{\sim}{v}_j$ can be suppressed by making the force vector $\underset{\sim}{p}$ orthogonal to it, i.e.
$\underset{\sim}{v}_j^{\mathsf{T}} \underset{\sim}{p} = 0$. Alternatively the amplitude of mode $\underset{\sim}{v}_j$ can be increased by reducing the frequency differences $(\omega_j^2 - \omega^2)$. The receptance $\underset{\sim}{\alpha}_{rs}$ at station r due to a sinusoidal excitation at station s is obtained by inserting a force vector $\underset{\sim}{p}$ with a 1 in the s-th component and zeros elsewhere. If v_{rj} denotes the r-th component of vector $\underset{\sim}{v}_j$, it follows from (13) that

$$(14) \qquad \underset{\sim}{\alpha}_{rs} = \sum_{j=1}^{n} v_{rj} v_{sj} \frac{1}{m_j \left\{ (\omega_j^2 - \omega^2) + 2i\zeta_j \omega_j \omega \right\}} .$$

A similar expression for the receptance of a continuous system can be derived. For instance, in the case of a uniform beam of visco-elastic material we have

$$\alpha_{rs} = \sum_{j=1}^{\infty} W_j(x_r) W_j(x_s) \frac{1}{M_j \left\{ (\omega^2 - \omega^2) + 2i\zeta_j \omega_j \omega \right\}} .$$

where the modal functions $W_j(x)$ satisfy the equations

$$EI W_r^{(4)}(x) - m\omega_r^2 W_r(x) = 0$$

and the end conditions on the beam. The mass per unit length of the beam is denoted by m . The generalized mass is defined as $M_j = \int_0^{\ell} m W_j^2(x)\,dx$. The modal damping factor is defined as $\zeta_j = V\omega_j / 2E$ where V and E are the viscous and elastic modduli in the stress–strain relation $\sigma = E\varepsilon + V\dot{\varepsilon}$.

All modes above a certain frequency will be over–critical damped, since the damping factor increases with ω_j ; this is characteristic of all systems in which the damping matrix is proportional to the stiffness matrix.

8.3. Normal coordinates

The n normal coordinates of system (1) are related to the initial more or less arbitrarily chosen coordinates $\underset{\sim}{q}$ by the matrix transformation

$$\underset{\sim}{q} = \underset{\sim}{V}\underset{\sim}{\xi} = \xi_1\underset{\sim}{v}_1 + \ldots + \xi_n\underset{\sim}{v}_n , \tag{15}$$

where $\underset{\sim}{V}$ is composed of the n column vectors $\underset{\sim}{v}_j$, i.e.

$$\underset{\sim}{V} = \left[\underset{\sim}{v}_1, \underset{\sim}{v}_2, \ldots, \underset{\sim}{v}_n\right] = \begin{bmatrix} v_{11} & v_{12} & v_{13} & \ldots \\ v_{21} & & & \\ \vdots & & & \\ v_{n1} & & & \end{bmatrix}. \tag{16}$$

Substitution of (15) into (1) yields

$$\underset{\sim}{A}\underset{\sim}{V}\ddot{\underset{\sim}{\xi}} + \underset{\sim}{B}\underset{\sim}{V}\dot{\underset{\sim}{\xi}} + \underset{\sim}{C}\underset{\sim}{V}\underset{\sim}{\xi} = \underset{\sim}{Q} ,$$

which after prelumtiplying by $\underset{\sim}{V}^{T}$ and taking account of the orthogonality relations (8), (9), (10), (11) becomes

$$\ddot{\xi}_1 + 2\zeta_1\omega_1\dot{\xi}_1 + \omega_1^2\xi_1 = \frac{1}{m_1}\Xi_1$$

$$\ddot{\xi}_2 + 2\zeta_2\omega_2\dot{\xi}_2 + \omega_2^2\xi_2 = \frac{1}{m_2}\Xi_2$$

(17)
$$\cdots$$

$$\ddot{\xi}_n + 2\zeta_n\omega_n\dot{\xi}_n + \omega_n^2\xi_n = \frac{1}{m_n}\Xi_n$$

where the normal forces Ξ_j are the components of $\underset{\sim}{V}^T\underset{\sim}{Q}$. If condition (11) on the damping matrix were not fulfilled the normal equations (17) would be coupled through the velocity terms. However, since the current methods of determining the generalized masses m_j are based on an undamped model and the use of viscous damping is more of a convenient artifice than a true description of the usual mechanical situation, there seems to be little point in considering other than proportional damping except as a last resort.

It is seen from the second way of writing (15) that the <u>normal coordinates</u> ξ_j are a measure of what proportion of each normal mode appears in the response $\underset{\sim}{q}$.

Chapter 9 RANDOM VIBRATION RESPONSE II: SIMPLE EXAMPLES

9.1. System with two degrees of freedom

To illustrate the situation where a system is to be regarded as one with a few degrees of freedom, with lumped parameters, we consider the conventional system of Fig. 9.1 We suppose that the input consists of random movement of the base, represented by the coordinate z , and that we are interested in the resulting movements of the two masses represented by the coordinates x and y .

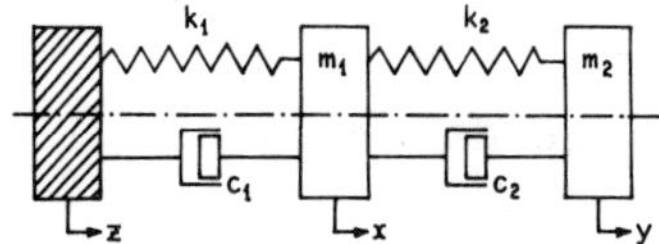

We require the receptances α_{xz} and α_{yz} , giving the outputs at x and y respectively for a unit sinusoidal input at z . The receptances here are ratios of displacements instead of ratios of displacements to force, but it is convenient to use the term in a wider sense to cover inputs and outputs of various natures.

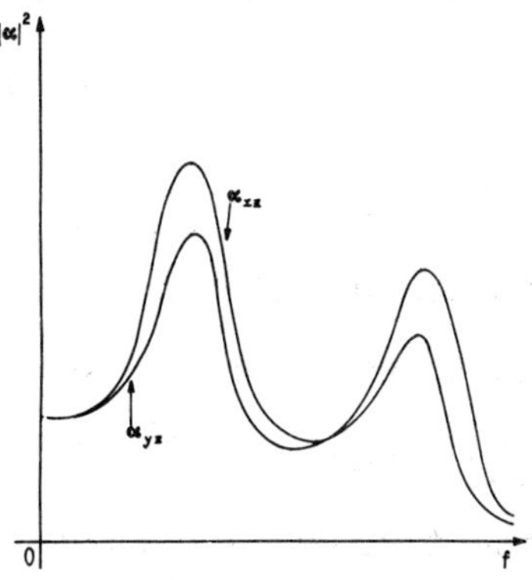

Fig. 9.1

The differential equations for this case are

$$m_1\ddot{x} + (c_1 + c_2)\dot{x} + (k_1 + k_2)x - c_2\dot{y} - k_2 y \;=\; c_1\dot{z} + k_1 z$$

and

$$m_2\ddot{y} + c_2\dot{y} + k_2 y - c_2\dot{x} - k_2 x = 0 .$$

For $z = z_0 e^{i\omega t}$, $x = x_0 e^{i\omega t}$, $y = y_0 e^{i\omega t}$, we get the desired ratios x_0/z_0 and y_0/z_0 as functions of ω:

$$\alpha_{xz}(i\omega) = \frac{k_1 k_2 + i\omega(k_1 c_2 + k_2 c_1) - \omega^2(m_2 k_1 + c_1 c_2) - i\omega^3 m_2 c_1}{D}$$

$$\alpha_{yz}(i\omega) = \frac{k_1 k_2 + i\omega(k_1 c_2 + k_2 c_1) - \omega^2 c_1 c_2}{D}$$

and the denominator D is

$$m_1 m_2 \omega^4 - i\omega^3(m_1 c_2 + m_2 c_1 + m_2 c_2) - \omega^2(m_1 k_2 + m_2 k_1 + m_2 k_2 + c_1 c_2) +$$
$$+ i\omega(k_1 c_2 + k_2 c_1) + k_1 k_2 .$$

If there is no damping, the denominator reduces to a quadratic function of ω^2, the roots of which give the two natural frequencies of the system. For moderate amounts of damping, the quantities $|\alpha_{xz}|^2$ and $|\alpha_{yz}|^2$ will both have peaks in the neighbourhood of these frequencies, the shape of the peaks depending on the parameters of the system. If the random input is white noise at z, the spectral densities at x and y will be proportional to these quantities. The mean square of the outputs can be found as before

$$<x^2(t)> = a^2 \int_0^\infty |\alpha(if)|^2 \, df , \text{ as in } 7.4$$

using the table of integrals referred to there. A detailed study

of this system, with reference to the effect of different values
of the parameters, is given in Crandall and Mark (Ref. 1.1).

9.2. Use of normal modes of vibration

The most general method of analysis for a system
with many degrees of freedom is that based on the concept of normal modes, as described in the last lecture. This method is best
illustrated in terms of a system with distributed parameters,
which we think of as having an infinite number of degrees of
freedom. We shall again use the transverse vibration of a beam
as a convenient illustration, since for this case the idea of the
modes of vibration is a familiar one, and the modes themselves
can be easily visualized.

We suppose then that we have a random force P
applied to the beams at a point determined by a coordinate x_1,
measured along the length, and that we wish to find the movement
at a point with coordinate x_2. It has been established that the
movement $w(x_2,t)$ at this point for a sinusoidal input $P_0 e^{i\omega t}$ is
given by

$$w(x_0,t) = \sum_{r=1}^{\infty} \frac{w_r(x_2)\, w_r(x_1)}{M_r(\omega_r^2 - \omega^2 + i2\zeta_r \omega_r \omega)} P_0 e^{i\omega t} = \alpha_{21}(i\omega) P_0 e^{i\omega t},$$

where $\alpha_{21}(i\omega)$ is the receptance linking output displacement at
x_2 with input force at x_1.

In this formula, $w_r(x)$ is the mode associated

with the natural frequency ω_r , and M_r is the generalised mass for this mode. ζ_r is the damping coefficient for viscous damping, which must be small.

The spectral density of the output is then given as usual by

$$S_w(f) \;=\; |\alpha(if)|^2 S_p \; .$$

9.3. Uniform simply supported beam

To illustrate the significance of the previous formula, we can apply it to the simplest case of a uniform beam simply supported at the ends. For this case the natural frequencies are

$$\omega_1^2 = \frac{\pi^4 EI}{m\ell^4} \; ; \quad \omega_2^2 = \frac{2^4\pi^4 EI}{m\ell^4} \; ; \quad \omega_3^2 = \frac{3^4\pi^4 EI}{m\ell^4} \quad \text{and so on,}$$

so that $\omega_2 = 2^2 \cdot \omega_1 \; ; \; \omega_3 = 3^2 \cdot \omega_1 \; .$

The corresponding modes of vibration are sinusoidal

$$w_1 = \sin\frac{\pi x}{\ell} \; ; \quad w_2 = \sin\frac{2\pi x}{\ell} \quad \text{and so on.}$$

The generalized mass M_r was found in the previous lecture to be

$$M_r \;=\; \int_0^\ell m w_r^2(x)\,dx \; , \quad \text{in which } m \text{ is now constant,}$$

$$=\; \int_0^\ell m \sin^2\frac{r\pi x}{\ell}\,dx$$

$$= \frac{1}{2} m\ell \quad \text{for each mode}$$

The orthogonality of the modes is in this case obvious from the orthogonal property of the sine functions.

We can now write down α_{21} as the sum of an infinite series of terms. We can put the damping coefficients zero meantime.

$$\alpha_{21} = \frac{2}{m\ell} \left\{ \frac{\sin\dfrac{\pi x_2}{\ell}\sin\dfrac{\pi x_1}{\ell}}{\omega_1^2 - \omega^2} + \frac{\sin\dfrac{2\pi x_2}{\ell}\sin\dfrac{2\pi x_1}{\ell}}{16\omega_1^2 - \omega^2} + \frac{\sin\dfrac{3\pi x_2}{\ell}\sin\dfrac{3\pi x_1}{\ell}}{81\omega_1^2 - \omega^2} + \ldots \right\} .$$

It will be noted that in expressions of this sort some of the numerators may be zero; for example, if either x_1 or x_2 were equal to $\frac{1}{3}\ell$, the terms $\dfrac{3\pi x}{\ell}$ would be zero, and so would higher multiples of three. These zero terms represent modes which have a node at the point of application of the force or at the point of displacement measurement.

We note that there will be peaks in $|\alpha|^2$ at all the natural frequencies of the system, except as just noted where the points x_1 or x_2 are at, or near, a node for the particular frequency. The complete expansion of $|\alpha|^2$ will be complicated by the occurrence of the product terms from the series.

We can make a rough approximation by noting that

if the frequencies are widely spaced (as they are in our example),
the values of $|\alpha|^2$ in the neighbourhood of any one peak will de-
pend in the main on one term only in the series for α, for example in the neighbourhood of the second frequency ω_2 we can represent the peak (if there is one) by the term

$$
|\alpha_{21}|^2 = \frac{4}{m\ell^2}\frac{\left(\sin\dfrac{2\pi x_2}{\ell}\sin\dfrac{2\pi x_1}{\ell}\right)^2}{\left(\omega_2^2 - \omega^2\right)^2} .
$$

On this basis we can make an estimate of the form of $|\alpha|^2$ near
each peak, including damping terms in each case.

If the damping action is assumed to be hysteretic
rather than viscous in nature, we start in the equation of motion
with a complex elastic coefficient, which we can write in the
form $k(1 + i\eta)$. This leads to the form $\left(\omega_r^2 - \omega^2 + i\eta_r\omega_r^2\right)$ in the
denominator for α . Since we use empirical coefficients for ζ
and η , it does not make much difference which form we use in
this sort of calculation.

The method of this paragraph is fully discussed
in Robson (Ref. 1.2).

9.4. Derivatives of random functions

We are frequently concerned with derivatives of
random functions with respect to the independent variable, for
example when $x(t)$ represents displacement and the time deriva-

The set of derivatives of the realisations of a stationary ergodic process form in general a similar randon process, and the concepts which we have developed can be applied to the derived process. In particular, if the original process has a Gaussian distribution of magnitude, so has the derivative.

We start with finding the derivative of the auto-correlation function $R_x(\tau)$ of $x(t)$ with respect to the time interval τ .

$$R_x(\tau) \;=\; <x(t)x(t+\tau)>$$

$$R_x(\tau+\delta\tau) \;=\; <x(t)x(t+\tau+\delta\tau)> \;.$$

Now $x(t+\tau+\delta\tau) = x(t+\tau) + x'(t+\tau)\delta\tau$,

where $x'(t+\tau)$ means $\dfrac{dx(t+\tau)}{d(t+\tau)}$; that is $\dot{x}$ at $t+\tau$

$$\therefore R_x(\tau+\delta\tau) \;=\; <x(t)x(t+\tau)> + <x(t)x'(t+\tau)>\delta\tau$$

and $R_x'(\tau) = \dfrac{dR_x}{d\tau} = <x(t)x'(t+\tau)> \;.$

This is an example of a cross-correlation function, referred to in § 2.11, measuring the correlation between $x(t)$ at any time and $\dot{x}(t)$ at an interval τ later, so we could write: $R_x'(\tau) = R_{x\dot{x}}(\tau)$, using a double suffix notation.

We can also write

$$R_x(\tau) \;=\; <x(t)x(t-\tau)>$$

$$R_x(\tau + \delta\tau) \;=\; <x(t)x(t - \tau - \delta\tau)>$$

and $x(t - \tau - \delta\tau) = x(t - \tau) - x'(t - \tau)\delta\tau$, the negative sign occur-

ring since $\dfrac{d(t - \tau)}{d\tau} = -1$

$$\therefore R_x'(\tau) \;=\; -<x(t)x'(t - \tau)> \;=\; -R_{x\dot{x}}(-\tau) \;.$$

Comparing these two results, we see that $R_x'(\tau)$ is an odd function. It follows that if it has a unique value $R_x'(0)$, it must be zero; but there will not be such a value if $R_x'(\tau)$ has a jump at the origin, as for example in the square wave of § 6.5, Fig. 6.2.

We can also write $R_x'(\tau) = <x(t-\tau)\cdot x'(t)>$ or $-<x(t+\tau)\cdot x'(t)>$. Let us obtain the second derivative from the first of these forms; a similar argument gives

$$R_x''(\tau) \;=\; -<x'(t)x'(t - \tau)> \;=\; -<x'(t)x'(t + \tau)> \;=\; -R_{\dot{x}}'(\tau),$$

which is the auto-correlation function for $\dot{x}(t)$.
If $R_x''(0)$ exists, it has the value $-R_{\dot{x}}(0) = -<\dot{x}(t)^2>$.
Continuing this argument we find

$$R_x'''(0) \;=\; 0; \quad R_x^{IV}(0) \;=\; +<\ddot{x}(t)^2>, \text{ and so on}$$

so the mean square of any derivative of $x(t)$ can be found from the derivatives of $R(\tau)$ at $\tau = 0$.

Turning now to the spectral densities

$$R_x(\tau) \;=\; \int_{-\infty}^{\infty} \frac{1}{2} S_x(f) e^{i2\pi f\tau} df \;=\; \int_0^{\infty} S_x(f) \cos 2\pi f\tau\, df \;.$$

Taking derivatives under the integral with respect to τ ,

$$R_x'(\tau) \;=\; -\int_0^{\infty} 2\pi f\, S_x(f) \sin 2\pi f\tau\, df$$

and

$$R_x''(\tau) \;=\; -\int_0^{\infty} 4\pi^2 f^2 S_x(f) \cos 2\pi f\tau\, df$$

so

$$\langle \dot{x}(t)^2 \rangle \;=\; -R_x''(0) \;=\; \int_0^{\infty} 4\pi^2 f^2 S_x(f)\, df$$

and

$$\langle \ddot{x}(t)^2 \rangle \;=\; +R_x^{IV}(0) \;=\; \int_0^{\infty} 16\pi^4 f^4 S_x(f)\, df \;.$$

We can thus obtain the mean square value of the derivatives from the spectral density.

We can obtain the spectral density of the derivative, using the result that $R_{\dot{x}}(\tau) = -R_x''(\tau)$

$$S_{\dot{x}}(f) \;=\; 2\int_{-\infty}^{\infty} R_{\dot{x}}(\tau) e^{-i2\pi f\tau}\, d\tau$$

$$=\; -2\int_{-\infty}^{\infty} R_x''(\tau) e^{-i2\pi f\tau}\, d\tau$$

$$=\; 4\pi^2 f^2 S_x(f)$$

and similarly for higher derivatives.

Chapter 10

RANDOM VIBRATION RESPONSE III : CROSS–CORRELATION

10.1. Cross-correlation

Consider a random process $\{z(t)\}$ made up by the addition of two random processes $\{x(t)\}$ and $\{y(t)\}$, so that

(1)
$$\{z(t)\} = \{x(t)\} + \{y(t)\} .$$

Let the autocorrelation functions be $R_z(\tau), R_x(\tau), R_y(\tau)$ respectively.

$$\text{Then } R_z(\tau) = \,<z(t)\,z(t+\tau)>$$

$$= \,<[x(t)+y(t)][x(t+\tau)+y(t+\tau)]>$$

(2)
$$= <x(t)\,x(t+\tau)> + <x(t)\,y(t+\tau)> + <y(t)x(t+\tau)> + <y(t)y(t+\tau)> .$$

The first and fourth terms are simply $R_x(\tau)$ and $R_y(\tau)$: the second and third terms are new to us and must be given new names. They are cross–correlation functions, and denoting them by $R_{xy}(\tau)$ and $R_{yx}(\tau)$ we can write (2) as

(3)
$$R_z(\tau) = R_x(\tau) + R_{xy}(\tau) + R_{yx}(\tau) + R_y(\tau) .$$

Note that $R_z(\tau)$ cannot be found if only $R_x(\tau)$ and $R_y(\tau)$ are given. We must also be given the two cross–correlation functions which allow for any connection that may exist between $x(t)$ and $y(t)$. If for example there is no connection they will both be zero and can be ignored: if $x(t) = y(t)$ it follows from (3) that $R_z(\tau) = 4R_x(\tau)$, as would be expected.

It will be convenient to define also <u>cross spectral densities</u> by simply taking Fourier transform of the cross-correlation functions. Thus

$$
\begin{aligned}
S_{xy}(f) &= 2\int_{-\infty}^{\infty} R_{xy}(\tau)\,e^{-i2\pi f\tau}\,d\tau\ , \\
S_{yx}(f) &= 2\int_{-\infty}^{\infty} R_{yx}(\tau)\,e^{-i2\pi f\tau}\,d\tau\ .
\end{aligned}
\tag{4}
$$

Then taking the Fourier transform of both sides of (3) we have

$$
S_z(f) \;=\; S_x(f) + S_{xy}(f) + S_{yx}(f) + S_y(f)\ .
\tag{5}
$$

It may be noted that the cross terms, themselves in general complex, will be conjugate.

Whenever we are considering the interaction of a number of connected quantities their cross-correlations must be allowed for. Where, for example, a system is subject to multiple excitations we must include the effect of the cross-correlations of the excitations in determining the response.

The use of the double subscript for the cross quantities, as in $R_{xy}(\tau)$, suggests the propriety of using it also for the direct quantities e.g. $R_{xx}(\tau)$. This is in fact often done, but we shall retain our original notation.

10.2. Response to two forces

Let a structure be excited by two randomly varying forces $P(t)$ and $Q(t)$, as shown, having spectral densities $S_P(f)$,

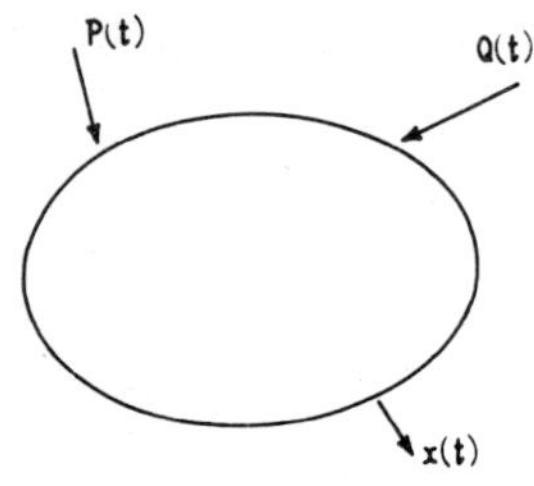

$S_Q(f)$, and cross spectral densities $S_{PQ}(f)$, $S_{QP}(f)$. Then the spectral density $S_x(f)$ of the response $x(t)$ at a given point can be shown – by suitable extension of the single-force analysis – to be given by

$$(6) \qquad S_x = \alpha_{xP}^* \alpha_{xP} S_P + \alpha_{xP}^* \alpha_{xQ} S_{PQ} + \alpha_{xQ}^* \alpha_{xP} S_{QP} + \alpha_{xQ}^* \alpha_{xQ} S_Q ,$$

where α_{xP} is the receptance connecting harmonically varying displacement x with harmonically varying force P , and $*$ indicates complex conjugate. Again the cross-correlation terms may be important: again the second and third terms will be complex conjugate: again if there is no correlation between $P(t)$ and $Q(t)$ these terms may be ignored. For $Q(t)=0$, (6) reduces to the expected result.

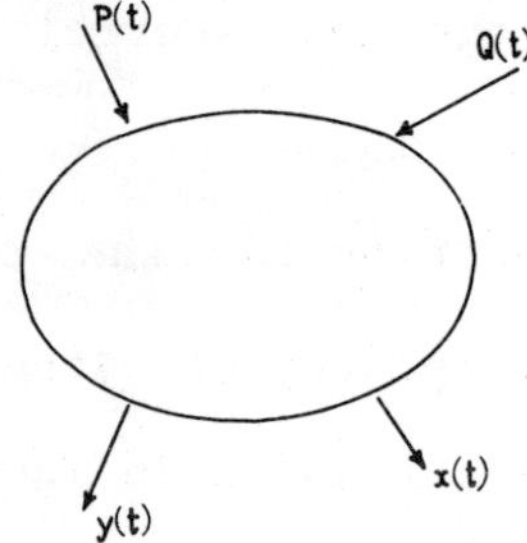

We may sometimes be interested in combinations of two different displacements $x(t)$ and $y(t)$ – as for example if we are interested in relative motion of two points. An expression for $S_{xy}(f)$ can be obtained corresponding to (6)

$$(7) \quad S_{xy} = \alpha_{xP}^* \alpha_{yP} S_P + \alpha_{xP}^* \alpha_{yQ} S_{PQ} + \alpha_{xQ}^* \alpha_{yP} S_{QP} + \alpha_{xQ}^* \alpha_{yQ} S_Q . \qquad (7)$$

If y is replaced by x throughout this reduces to (6).

It may be noted that even where only a single force acts, cross-correlation will still exist between the two responses $x(t)$ and $y(t)$.

10.3. Response to multiple excitation

When there are many applied forces $P_1, P_2, \ldots, P_n$ the spectral density of the response again contains contributions from all their direct spectral densities and from all the cross spectral densities. As quantities can only "cross" in pairs the result obtained is completely analogous to (7), in fact:-

$$S_{xy} = \sum_{r=1}^{n} \sum_{s=1}^{n} \alpha^*_{xP_r} \, \alpha_{yP_s} \, S_{P_r P_s} \; . \tag{8}$$

10.4. Response to distributed excitation

The general problem, in which the whole surface of a body is subjected to pressure, variable with respect to time and location, is complex in formulation (and in fact) though not in principle. The principles can be illustrated adequately by consider-ing a beam subjected to random transverse pressures over its length.

Consider a beam of length ℓ , subject to loading of intensity $p(x,t)$ per unit length. It will be convenient to investigate the displacement at the point x_1 due to pressures acting at points x_A and x_B . (Our basic problem must include loading at a pair of points so that when we integrate cross-correlation effects will be included).

As a condition of solution the cross spectral density $S^P(x_A, x_B; f)$ of the pressure must be known for all x_A, x_B . This can be determined by transformation of the cross-correlation function $R^P(x_A, x_B; \tau) = <p(x_A, t)p(x_B, t + \tau)>$.

Equation (8) will be applicable if we replace $P_r(t)$, $P_s(t)$ by $p(x_A, t)dx_A$, $p(x_B, t)dx_B$, and replace the summation signs by integrals. The spectral density $S^w(x_1; f)$ of the displacement $w(x_1, t)$ is then given by

$$(9) \qquad S^w(x_1; f) = \int_0^\ell \int_0^\ell \alpha_{1A}^* \alpha_{1B} S^P(x_A, x_B; f) dx_A dx_B .$$

10.5. Response in terms of normal modes

Although an expression such as (9) offers a formally complete solution, it is useful to develop it by expressing the receptances α_{1A} , α_{1B} in terms of the normal modes $w_r(x)$ and natural frequencies f_r of the system.

The receptance α_{1A} can be expressed as

$$(10) \qquad \alpha_{1A} = \sum_r \frac{w_r(x_1)w_r(x_A)}{4\pi^2 M_r(f_r^2 - f^2 + i2\zeta_r f_r f)} ,$$

where $M_r = \int_0^\ell w_r^2(x)\,m\,dx$, m being mass per unit length, and ζ_r is the viscous damping coefficient in the rth normal mode.

Substituting in (9) we have

$$S^w(x_1,f) = \sum_r \sum_s \left[\frac{w_r(x_1)w_s(x_1)}{16\pi^4 M_r M_s(f_r^2 - f^2 - i2\zeta_r f_r f)(f_s^2 - f^2 + i2\zeta_s f_s f)} \times \right.$$
$$\left. \int_0^\ell \int_0^\ell w_r(x_A)w_s(x_B)S^P(x_A,x_B;f)\,dx_A\,dx_B \right]. \tag{11}$$

If damping is small – as is common in vibration problems – and if response peaks are well separated, this result can be greatly simplified, for then the product terms $(r \neq s)$ may be neglected in comparison with the square terms $(r = s)$. Equation (11) then becomes

$$S^w(x_1,f) = \sum_r \left[\frac{w_r^2(x_1)}{16\pi^4 M_r^2\left[(f_r^2 - f^2)^2 + 4\zeta_r^2 f_r^2 f^2\right]} \times \right.$$
$$\left. \int_0^\ell \int_0^\ell w_r(x_A)w_r(x_B)S^P(x_A,x_B;f)\,dx_A\,dx_B \right]. \tag{12}$$

The spectrum will obviously have peaks near each of the natural frequencies f_r and at each of these a single term of the summation will predominate.

We have therefore – for these special but common conditions – effectively reduced the response at any frequency to the contribution of a single normal mode, and have only then a single-degree-of-freedom problem to consider.

Bibliography for Chapter 10

10.1   Robson J.D.: Introduction to Random Vibration:
           Edinburgh University Press 1963.

See also:

10.2   Powell A.: Chapter 8 of Crandall S.H. (ed) Random
           Vibration
           Technology Press/Wiley, 1958.

10.3   Bendat J.S. and Piersol A.G.: Measurement and Analysis
           of Random Data:
           Wiley, 1966.

10.4   Lin Y.K.: Probabilistic Theory of Structural Dynamics:
           McGraw-Hill 1967.

Chapter 11   APPLICATIONS I : VEHICLE RESPONSE

11.1. Introduction

When a vehicle travels along a road the undulations of the road-surface impose continuously varying displacements at the points of contact of tyres and road, and so give rise to responses in the form of stresses or acceleration in the various components of the vehicle.

If these undulations were to be represented by a deterministic form e.g. sinusoidal input, then it would be relatively easy using classical vibration techniques to calculate the vehicle response. However, road surface profiles are in general random in nature and we must seek other means of analysing the vibrations set up in the vehicle.

If we make the assumption that the random function which represents the road profile irregularities is a member function of an ergodic process, then we can develop the theory outlined in the previous lectures to enable us to claculate and predict vehicle response to road profile excitation.

11.2. Response to profile excitation

The road profile $y(x)$ is conveniently measured as a function of distance x , and if we assume that $y(x)$ is plotted about a zero mean we can define the autocorrelation function

$$(1) \qquad \mathcal{R}(\delta) = \langle y(x) \cdot y(x + \delta) \rangle$$

as a function of the lag δ . Correspondingly we can define the power spectral density of the road to be

$$(2) \qquad \mathcal{G}(n) = 2 \int_{-\infty}^{\infty} \mathcal{R}(\delta) e^{-i2\pi n\delta} d\delta$$

where n is the wave number or spatial frequency.

For a vehicle travelling at constant velocity v, the input to the vehicle due to the surface irregularities is seen as a function of time, $y(t)$. Since $x = vt$, then

$$(3) \qquad y(t) = y(vt) .$$

The spectral density $S^{i}(f)$ of $y(t)$ is related to $\mathcal{G}(n)$ since, combining (1) and (2) we get

$$
\begin{aligned}
\mathcal{G}(n) &= 2 \int_{-\infty}^{\infty} \langle y(x) \cdot y(x + \delta) \rangle e^{-i2\pi\delta n} d\delta \\
(4) \\
&= 2 \int_{-\infty}^{\infty} \langle y(vt) \cdot y(vt + v\tau) \rangle e^{-i2\pi\frac{f}{v}v\tau} v d\tau
\end{aligned}
$$

as $\quad \delta = v\tau , \quad \tau = t_2 - t_1 , \quad d\delta = v d\tau , \quad f = vn$

$$= 2v \int_{-\infty}^{\infty} <y(t)y(t+\tau)> e^{-i2\pi f\tau} d\tau$$

$$= vS^i(f) \tag{5}$$

$$\therefore S^i(f) = \frac{1}{v} \mathcal{S}(n) .$$

The vehicle response to a single input can therefore be obtained from the equation

$$S^0(f) = |\alpha(if)|^2 S^i(f) \tag{6}$$

if we can find a suitable expression for the receptance $\alpha(if)$.

11.3. Simple model. Single input

In Chapter 4 the receptance has been introduced as that quantity which gives the response of a system to a complex force of unit modulus and proportional to $e^{i\omega t}$.

In the case of a vehicle traversing a surface profile, this profile imposes a displacement (or displacements) on the system. Thus we must widen the definition of receptance to include – the response of a system due to a displacement of unit modulus and proportional to $e^{i\omega t}$. With this latter definition we can proceed to evaluate the receptance for a given vehicle model. For example for the model in fig. 1 the equation of motion for the mass m due to the imposed vertical

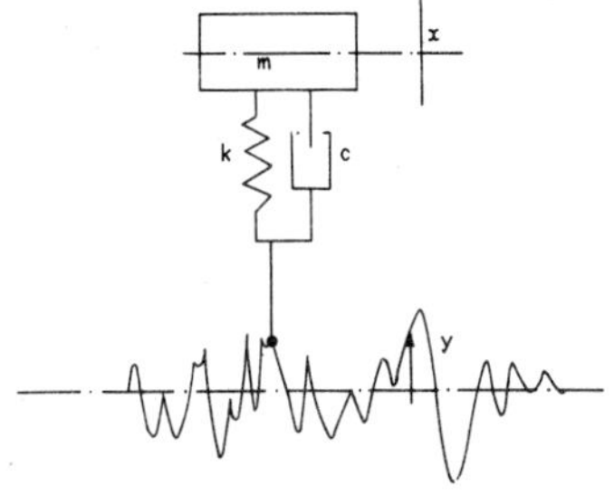

Fig. 1

displacement $y(t)$ arising from the random function $y(x)$ is

$$(7) \qquad m\ddot{x} + c\dot{x} + kx = c\dot{y} + ky$$

which with usual notation can be written as

$$(8) \qquad \ddot{x} + 2\zeta\omega_n\dot{x} + \omega_n^2 x = 2\zeta\omega_n\dot{y} + \omega_n^2 y$$

if $y = e^{i\omega t}$, then letting $x = X e^{i\omega t}$ we have

$$(9) \qquad
\begin{cases}
X = \dfrac{\omega_n^2 + i2\zeta\omega\omega_n}{\omega_n^2 - \omega^2 + i2\zeta\omega\omega_n} \\[3mm]
\text{or} \\[3mm]
X = \dfrac{f_n^2 + i2\zeta f f_n}{f_n^2 - f^2 + i2\zeta f f_n}
\end{cases}$$

where f_n is the natural frequency of the system. If we define the receptance $\alpha(x, y, f)$ to be the receptance of the displacement of coordinate x, due to the displacement of coordinate y in terms of the parameter f, then

$$(10) \qquad \alpha(x,y,f) = \frac{f_n^2 + i2\zeta f f_n}{f_n^2 - f^2 + i2\zeta f f_n}$$

It follows that

$$(11) \qquad \alpha(\ddot{x},y,f) = -4\pi^2 f^2 \alpha(x,y,f)$$

$$\alpha(x-y,y,f) = \frac{f^2}{f_n^2 - f^2 + i2\zeta ff_n} \quad . \tag{12}$$

Using a similar notation for the response spectrum we can establish this for the model shown. Combining equations (5), (6) and (10) we have

$$S_v^0(x,f) = \frac{1}{v}\left|\alpha(x,y,f)\right|^2 \mathcal{S}(n) \tag{13}$$

where subscript v denotes velocity dependence and $\underline{f = vn}$. Thus

$$S_v^0(x,f) = \frac{1}{v}\frac{f_n^4 + 4\zeta^2 f^2 f_n^2}{(f_n^2 - f^2)^2 + 4\zeta^2 f^2 f_n^2}\,\mathcal{S}(n) \quad . \tag{14}$$

The mean square value of the response can therefore be evaluated and is given by

$$\sigma_x^2 = \frac{1}{v}\int_0^\infty \frac{f_n^4 + 4\zeta^2 f^2 f_n^2}{(f_n^2 - f^2)^2 + 4\zeta^2 f^2 f_n^2}\,\mathcal{S}\!\left(\frac{f}{v}\right)df \quad . \tag{15}$$

11.4. Two input problem

It may be unrealistic to consider a vehicle to have only one displacement imposed by the surface profile and we shall now consider a vehicle having two inputs to the vehicle, one through the front suspension and one through the rear.

The general equation relating the response spec-

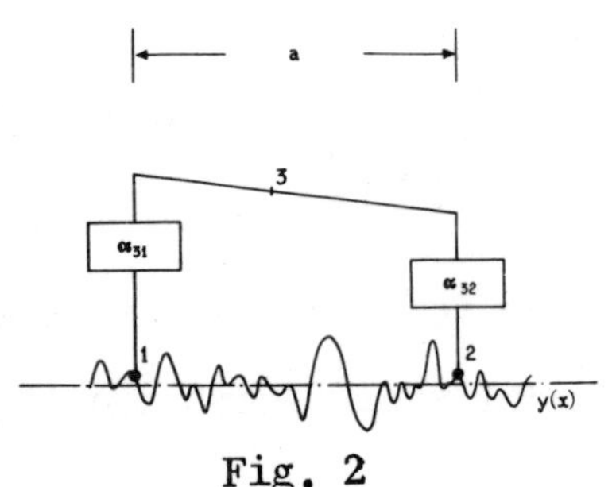

Fig. 2

trum of a point (3) (see fig. 2) due to two imposed displacements at (1) and (2) is

$$(16) \quad S_3^0(f) = \alpha_{31}\alpha_{31}^* S_1^i(f) + \alpha_{31}^*\alpha_{32} S_{12}^i(f) + \alpha_{31}\alpha_{32}^* S_{21}^i(f) + \alpha_{32}\alpha_{32}^* S_2^i(f) \; .$$

We assume that the points of contact follow the same profile, but separated by constant lag, and if $y_1(x)$ and $y_2(x)$ are the imposed displacements then

$$y_1(x) = y(x)$$
$$y_2(x) = y(x-a)$$

with $x = vt$ we have

$$R_{12}(\tau) = \langle y_1(t) \cdot y_2(t+\tau) \rangle$$

$$= \langle y_1(vt) \cdot y_2(vt+v\tau) \rangle$$

$$(17) \qquad = \langle y_1(x) \cdot y_2(x+\delta) \rangle$$

$$= \langle y(x) \cdot y(x-a+\delta) \rangle$$

$$= \mathcal{R}(\delta-a) \; .$$

Using the transform relationship

$$S_{12}^{i}(f) = 2\int_{-\infty}^{\infty} R_{12}(\tau)e^{-i2\pi f\tau}\,d\tau$$

$$= 2\int_{-\infty}^{\infty} \mathcal{R}(\delta-a)e^{-i2\pi vn\frac{\delta}{v}}\,d\left(\frac{\delta}{v}\right)$$

as $\quad f = vn \quad$ and $\quad \tau = \delta/v$

$$S_{12}^{i}(f) = \frac{2}{v}\int_{-\infty}^{\infty} \mathcal{R}(\delta-a)e^{-i2\pi\delta n}\,d\delta$$

$$= \frac{2}{v}\int_{-\infty}^{\infty} \mathcal{R}(\delta-a)e^{-i2\pi n(\delta-a)}\,d(\delta-a)\cdot e^{-i2\pi na} \qquad (18)$$

i.e. $\quad S_{12}^{i}(f) = \frac{1}{v}e^{-i2\pi na}\,\mathcal{S}(n)\,.$

The response relationship now becomes (omitting subscript (3))

$$s^{0}(f) = \left[\alpha_1\alpha_1^* + \alpha_2\alpha_2^* + \alpha_1\alpha_2^* e^{i2\pi na} + \alpha_1^*\alpha_2 e^{-i2\pi na}\right]\frac{1}{v}\,\mathcal{S}(n)\,. \qquad (19)$$

This can be written as

$$s^{0}(f) = \frac{1}{v}\left[A(f) + B(f)\cos 2\pi na + C(f)\sin 2\pi na\right]\mathcal{S}(n) \qquad (20)$$

where

$$A(f) = \alpha_1\alpha_1^* + \alpha_2\alpha_2^*$$

$$B(f) = 2\,\mathcal{R}e\left[\alpha_1^*\alpha_2\right]$$

$$C(f) = 2\,\mathcal{G}m\left[\alpha_1^*\alpha_2\right]\,.$$

Equation (20) can now form the basis for calculation fo the vehicle response spectrum due to random profile excitation, provided that the correlation between near side and off-side is unimportant.

11.5. Calculation of response — practical example

Figure 3 shows typical road profile spectra. They have been abstracted from various sources (see Bibliography (2) to (5)). Let us consider the profile spectrum of one of these routes, (fig. 4) and illustrate how the techniques of random vibration analysis can be used to aid the design engineer.

Suppose we wish to obtain the r.m.s. stress in the front spring of a vehicle on the road whose spectrum is given in figure 4. Then, if we assume that we can simulate the front near side suspension system of the car by the model of figure 5 we can allocate to the parameters of the model, values derived from the design parameters of the vehicle, or as in the case of this example, parameters based on those of an actual vehicle.

The response spectrum for the stress in the coil spring can be calculated from

$$S_v^0(z_{s_1} - z_1, f) = \left| \alpha(z_{s_1} - z_1, y, f) \right|^2 \frac{1}{v} \mathcal{S}(f/v)$$

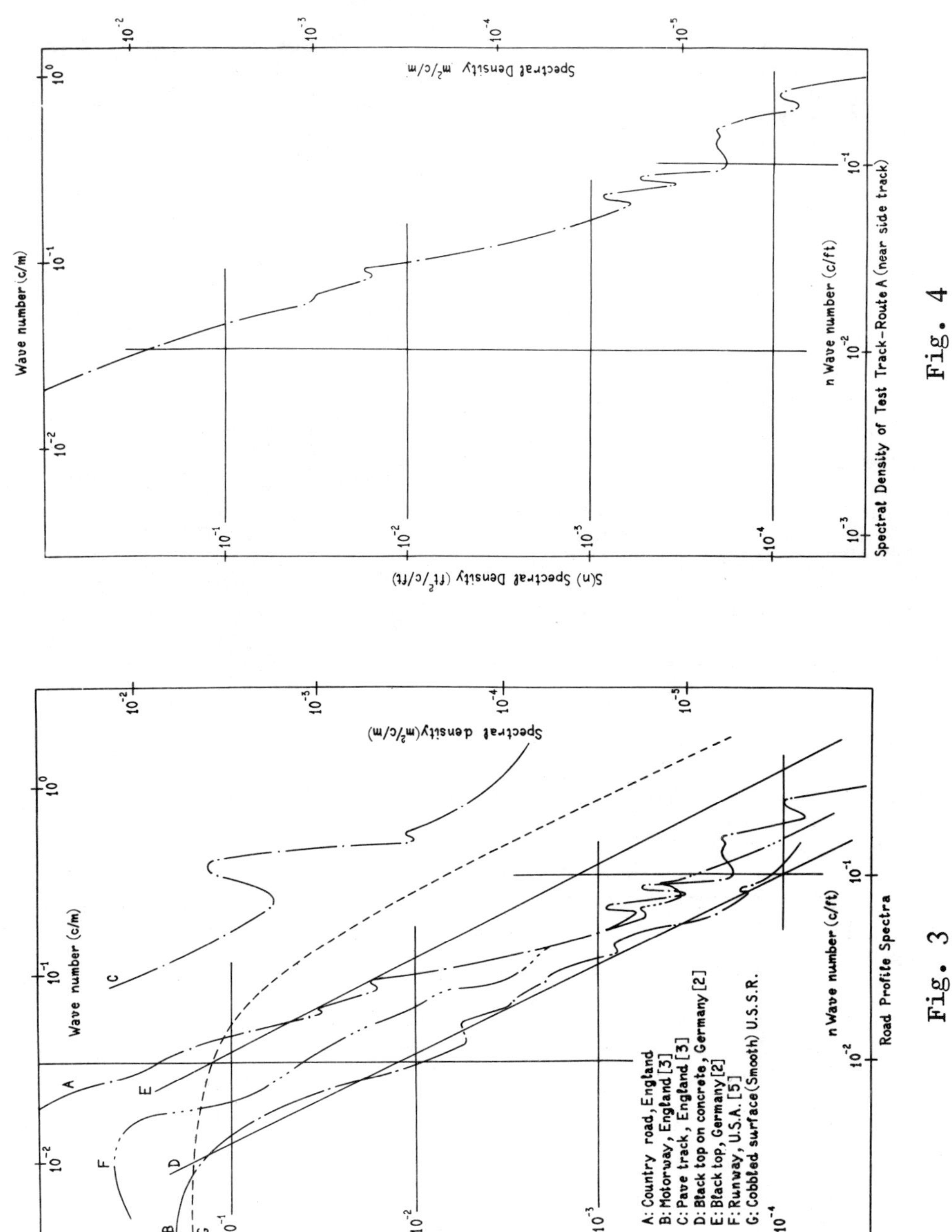
Wave number (c/m)
Spectral Density m²/c/m
S(n) Spectral Density (ft²/c/ft)
n Wave number (c/ft)
Spectral Density of Test Track – Route A (near side track)
Fig. 4
Wave number (c/m)
Spectral density (m²/c/m)
S,n Spectral Density (ft²/c/ft)
n Wave number (c/ft)
Road Profile Spectra
Fig. 3
A: Country road, England
B: Motorway, England [3]
C: Pave track, England [3]
D: Black top on concrete, Germany [2]
E: Black top, Germany [2]
F: Runway, U.S.A. [5]
G: Cobbled surface (Smooth) U.S.S.R.

and this is shown in figure 6 for a given speed v , together with the experimental response spectrum obtained from analysis of a strain-gauge record for the actual spring while the vehicle travelled along the test route.

The mean square value of the response can be evaluated and if the probability density function is known, the peak values obtained. Improved predicted results can be obtained if more sophisticated models are used (see Bibliography reference 6).

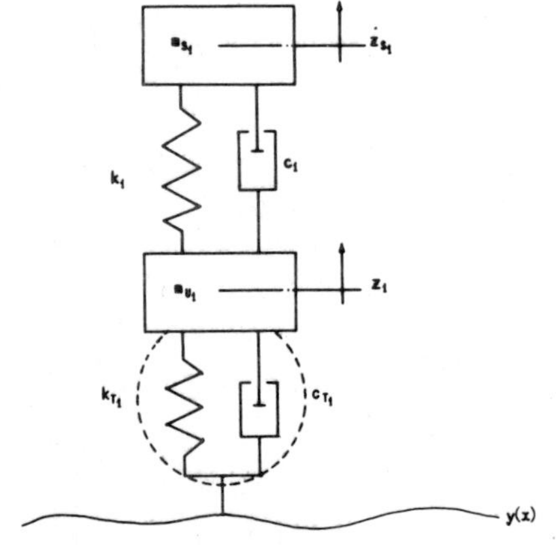

Single input, 2-degree of freedom, linear vehicle model
(front suspension)   Fig. 5

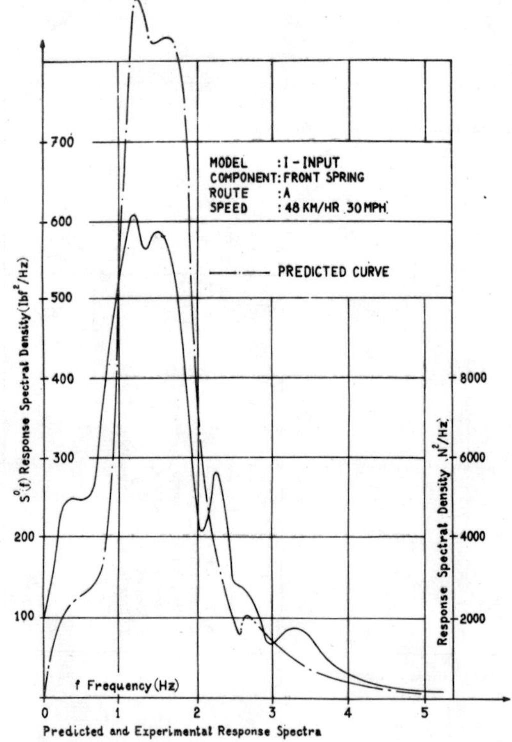

Fig. 6

Bibliography for Chapter 11

11.1   Van Deusen B.D.:   A Study of the Vehicle Ride Dynamics
        Aspect of Ground Mobility
        MERS Project Final Report, Vol. 1, Summary, U.S.
        Army Corps of Engineers, Waterways Experiment
        Station, Contractor Report No. 3-114, May 1965.

This paper gives an appraisal of the subject to May 1965
and includes a large Bibliography.

11.2   Braun H.:   Untersuchungen Über Fahrbahnunebenheiten
        (Investigation of Road Surface Roughness)
        Deut. Kraft., No. 186, p. 83, 1966.

This article describes a device for measuring road surface
profile and presents the spectral densities for various types of
roads in West Germany.

11.3   Motor Industry Research Association, Nuneaton, England:
        Report 1970/5: Measurement and Analysis of Road
        Surface Roughness.

11.4   Pevzner Y.M., Tikhonov A.A.:   An investigation into the
        statistical properties of the Microprofile of the
        Main types of Motor Road.
        Avtom. Prom 1 pp. 15-18, 1964.

11.5   Walls J.H., Houbolt J.C., Press H.:   Some Measurements
        and Power Spectra of Runway Roughness.
        N.A.C.A. T.N. 3305.

11.6   Dodds C.J., Robson J.D.:   The response of vehicle com-
        ponents to random road surface undulations.
        13th FISITA Congress, paper 17.2, June 1970.

Chapter 12

DESCRIPTION IV : FURTHER TOPICS IN PROBABILITY

12.1. Characteristic functions of probability distributions

Given a random variable having a probability density function $p(x)$, we introduce the Fourier transform $\Phi(u)$:

$$\Phi(u) = \int_{-\infty}^{\infty} e^{-ixu} p(x) \cdot dx \, , \quad \text{and conversely} \quad p(x) = \frac{1}{2\pi} \int_{-\infty}^{\infty} e^{ixu} \Phi(u) \cdot du.$$

We write the transform and the inverse in the more usual form with the factor 2π , since the interpretation of u as a frequency is meaningless here; some authors, however, interchange the signs of the exponent in these formulae.

This function $\Phi(u)$ is called the 'characteristic function' of the probability distribution of x ; mathematically it defines the nature of the distribution just as well as $p(x)$, and in theoretical work it is often easier to handle.

We note first that: $\Phi(0) = \int_{\infty}^{-\infty} p(x) \cdot dx = 1$,

and $\qquad \Phi'(u) = \dfrac{d\Phi}{du} = \int_{-\infty}^{\infty} -ixe^{-ixu} p(x) \cdot dx$

$\therefore \Phi'(0) = -i \int x \cdot p(x) \cdot dx = -iE[x] \, , \quad \text{and} \quad E[x] = i\Phi'(0) .$

As defined in § 2.3, $E[x]$ is the 'expectation' of x ; that is, the mean value of x to be expected from a large number of occasions.

Similarly we find:
$$E[x^2] = i^2 \Phi''(0)$$
$$E[x^n] = i^n \Phi''(0)$$

where $E[x^n]$ is the n'th moment of the distribution.

If now we suppose $\Phi(u)$ to be expanded in a Taylor series of powers of u ,

$$\Phi(u) = \Phi(0) + \Phi'(0)\cdot u + \Phi''(0)\cdot\frac{u^2}{\underline{|2}} + \ldots$$
$$= 1 + \frac{E[x]}{i}\cdot u + \frac{E[x^2]}{i^2}\cdot\frac{u^2}{\underline{|2}} + \ldots .$$

Hence if we obtain this series in any way, we can read off the moments from the coefficients in the series. For example, with the Gaussian distribution with mean $\bar{x}$ and variance σ^2,

$$p(x) = \frac{1}{\sigma\sqrt{2\pi}} e^{-\frac{(x-\bar{x})^2}{2\sigma^2}} .$$

The Fourier transform is $\Phi(x) = e^{-(i\bar{x}u + \frac{1}{2}\sigma^2 u^2)}$, from the table given in Chapter 6. Expanding in powers of u gives:

$$\Phi(x) = 1 - i\bar{x}u - (\sigma^2 + \bar{x}^2)\frac{u^2}{\underline{|2}} + (3i\bar{x}\sigma^2 + i\bar{x}^3)\frac{u^3}{\underline{|3}} + (3\sigma^4 + \bar{x}^4)\frac{u^4}{\underline{|4}} + \ldots$$

so $E[x] = \bar{x}$; $E[x^2] = \sigma^2 + \bar{x}^2$; $E[x^3] = 3\bar{x}\sigma^2 + \bar{x}^3$; $E[x^4] = 3\sigma^4 + \bar{x}^4$; $\ldots$

12.2. Probability distribution of sum of two random functions

Given two random functions x and y with probability densities $p(x)$ and $q(y)$ respectively, we wish to find a density function $r(z)$, where $z = (x+y)$. We suppose here that x and y are independent; that is, $p(x)$ is not dependent on the value of y, nor $p(y)$ on the value of x.

If two events are independent, the probability that both occur is the product of their probabilities. So the probability of the occurrence of a specific value of $z = (x+y)$ can be considered as the probability of the occurrence of any value x_1 of x, in association with the value $(z - x_1)$ of y, summed over all values of x_1.

Thus $r(z) = \int_{-\infty}^{\infty} p(x_1) q(z - x_1) dx_1$ or alternatively $\int_{-\infty}^{\infty} p(z - y_1) q(y_1) dy_1$.

This type of integral is of frequent occurrence and $r(z)$ is called the 'convolution' of the functions p and q, written $p^* q$. (The term is applied in different contexts to integrals with different limits). It may be possible to evaluate this integral directly, but the difficulty is avoided by taking the Fourier transform.

$$
\Phi_r(u) = \int_{-\infty}^{\infty} r(z) e^{-izu} dz
$$

(a)
$$
= \int_{-\infty}^{\infty} \left\{ \int_{-\infty}^{\infty} p(x_1) \cdot q(z - x_1) \cdot dx_1 \right\} e^{-izu} dz
$$

$$= \int_{-\infty}^{\infty} p(x_1) e^{-ix_1 u}\, dx_1 \int_{-\infty}^{\infty} q(z - x_1) e^{-i(z - x_1)u}\, d(z - x_1)$$

(b)

$$= \Phi_p(u)\Phi_q(u) .$$

This is the standard result that the transform of the convolution is the product of the transforms. (See, for example, Kaplan, Ref. 3.2). In terms of probability density functions, this means that the characteristic function of the sum of two independent random variables is the product of the characteristic functions of the separate variables. This answers our problem.

For example if x and y have Gaussian distributions with different means and variances;

$$\Phi_x(u) = \exp\left[-i\bar{x}u - \frac{1}{2}\sigma x^2 u^2\right] ; \qquad \Phi_y(u) = \exp\left[-i\bar{y}u - \frac{1}{2}\sigma y^2 u^2\right]$$

$$\text{and}\quad \Phi_z(u) = \exp\left[-i(\bar{x} + \bar{y})u - \frac{1}{2}(\sigma x^2 + \sigma y^2)u^2\right]$$

so $r(z)$ has also a Gaussian distribution with mean $\bar{z} = \bar{x} + \bar{y}$, and variance $\sigma_z^2 = \sigma_x^2 + \sigma_y^2$. We can readily extend this to show that the form of the Gaussian distribution is preserved in any linear combination of random variables, as we have previously assumed.

The additive relation between means and variance is valid for any probability density functions, as may be verified from the series expansion used above, but the form of the distribution is not usually preserved.

The substance of these paragraphs is given in most books on probability theory, including those listed in Chapter 2.

12.3. Joint probability density

If we have two random variables x and y which are not independent, we must define a joint probability density $p(x,y)$, as a function of both x and y , such that

$$p(x_0,y_0)\delta x_0,\delta y_0 = Pr[x_0 < x < x_0 + \delta x_0; y_0 < y < y_0 + \delta y_0] .$$

We think now of our unit of probability spread over an $x-y$ surface to give for each element of area $\delta x \cdot \delta y$ the appropriate density $p(x,y)$. We must therefore have

$$\int_{-\infty}^{\infty} \int_{-\infty}^{\infty} p(x,y) \cdot dx \cdot dy = 1 .$$

We shall want a two-dimensional Gaussian distribution, allowing for the two variables to have different variances. If the two variables were independent, we could take a form of the type:

$\exp[-a^2 x^2 - b^2 y^2]$; but we want a form linking the x and y variation, say: $\exp[-a^2 x^2 - b^2 y^2 + c^2 xy]$. The parameters involved are as usual

$$\sigma_x^2 = E[x^2] ; \quad \sigma_y^2 = E[y^2]$$

and a new parameter $\sigma_{xy}^2 = E[xy]$, to measure the link between the two variables. A suitable form is then

$$p(x,y) \;=\; \frac{1}{2\pi\sigma_x\sigma_y\sqrt{1-r^2}}\cdot\exp\left\{-\frac{\dfrac{x^2}{\sigma_x^2}-2r\dfrac{xy}{\sigma_x\sigma_y}+\dfrac{y^2}{\sigma_y^2}}{2(1-r^2)}\right\}$$

where we have written r for $\dfrac{\sigma_{xy}^2}{\sigma_x\sigma_y}$.

Corresponding forms for any number of variables can be defined, but it is convenient to write the parameters as elements of a matrix M . Thus in three variables we require a form

$$p(x,y,z) \;=\; \frac{1}{\sqrt{8\pi^3\cdot\det M}}\exp\left[-\frac{1}{2\det M}\left\{M_{xy}x^2 + M_{yy}y^2 + M_{zz}z^2 + \right.\right.$$
$$\left.\left. + 2M_{xy}xy + 2M_{yz}yz + 2M_{zx}zx\right\}\right] .$$

Here M is the 3×3 symmetrical matrix

$$\begin{bmatrix} \sigma_x^2 & \sigma_{xy}^2 & \sigma_{zx}^2 \\[2mm] \sigma_{xy}^2 & \sigma_y^2 & \sigma_{yz}^2 \\[2mm] \sigma_{zx}^2 & \sigma_{yz}^2 & \sigma_z^2 \end{bmatrix}$$

and the symbols M_{xy} denote the co-factor σ_{xy}^2 in the determinant of M : i.e. $M_{xy} = \sigma_{yz}^2\sigma_{zx}^2 - \sigma_{xy}^2\sigma_z^2$. A particular case which we shall use in Chapter 15 will be that with x as a function of time, as usual, and y and z representing the first and second derivatives $\dot{x}$ and $\ddot{x}$ respectively. We have previously found in § 9.4 that $\langle x(t)\cdot\dot{x}(t)\rangle = 0$, that is a variable and its first derivative are relatively independent. We can deduce at once that $\langle \dot{x}(t)\cdot\ddot{x}(t)\rangle$ is also zero; but the variable and its second derivative are not independent. In fact, $\langle x(t)\cdot\ddot{x}(t)\rangle = -\langle \dot{x}(t)^2\rangle$. This reduces our six possible parameters to three, since $\sigma_{x\dot{x}} = 0$; $\sigma_{\dot{x}\ddot{x}} = 0$; $\sigma^2_{x\ddot{x}} = -\sigma^2_{\dot{x}}$. So the matrix M becomes

$$
\begin{vmatrix}
\sigma^2_x & 0 & -\sigma^2_{\dot{x}} \\
0 & \sigma^2_{\dot{x}} & 0 \\
-\sigma^2_{\dot{x}} & 0 & \sigma^2_{\ddot{x}}
\end{vmatrix}
$$

The probability density function then becomes

$$
p(x,\dot{x},\ddot{x}) \;=\; \frac{1}{\sqrt{8\pi^3 \sigma^2_{\dot{x}}(\sigma^2_x\sigma^2_{\ddot{x}} - \sigma^4_{\dot{x}})}} \times
$$

$$
\exp\left[-\frac{\sigma^2_{\dot{x}}\sigma^2_{\ddot{x}}x^2 + (\sigma^2_x\sigma^2_{\ddot{x}} - \sigma^4_{\dot{x}})\dot{x}^2 + \sigma^2_x\sigma^2_{\dot{x}}\ddot{x}^2 + 2\sigma^4_{\dot{x}}x\ddot{x}}{2\sigma^2_{\dot{x}}(\sigma^2_x\sigma^2_{\ddot{x}} - \sigma^4_{\dot{x}})}\right].
$$

These results are discussed in Robson, Ref. 1.2, and in more detail in Bendat, Ref. 1.5.

12.4. Poisson probability distribution

This analysis deals with the situation where identical events occur in succession at random intervals of time, the average time between events being assumed known. We shall consider first the case of discrete events. Let us suppose that we perform an experiment on a number of occasions N , and that p is the probability of a certain result R , on each individual occasion. The probability that R will not occur on any one occasion is then $(1-p)$.

The probabilty that no result R will occur in N repetitions is $(1-p)^{N}$. The probability that precisely one result will occur in N repetitions is

$$Np(1-p)^{N-1} ;$$

and the probability that k results will occur is

$$\frac{\lfloor N}{\lfloor k \lfloor N-k} p^{k}(1-p)^{N-k} ;$$

the usual formula for the binomial coefficient.

Passing now to events on a continuous time basis, consider a total time T , which we divide into small intervals δt , and suppose that the probability of an event R in any one elementary interval δt is $\lambda \delta t$, λ being a constant probability density. In relation to the preceding notation, $\lambda \delta t$ corresponds to p , and $T/\delta t$ to N . We then represent a continuous process by taking the limit for $\delta t \to 0$ and $N \to \infty$. Then the probability of no event in time T is

$$(1-p)^N = (1-\lambda\cdot\delta t)^N = \left(1 - \frac{\lambda T}{N}\right)^N, \text{ and limit for } N \longrightarrow \infty \text{ is } e^{-\lambda T}.$$

Similarly the probability of one event is

$$Np(1-p)^{N-1} = \frac{T}{\delta t}\cdot\lambda\delta t\cdot\left(1 - \frac{\lambda T}{N}\right)^{N-1} \text{ and the limit is } \lambda T\cdot e^{-\lambda T}.$$

Finally the probability of k results is

$$\frac{\lfloor N}{\lfloor k \,\lfloor N-k}\,p^k(1-p)^{N-k} = \frac{N^k}{\lfloor k}\,p^k(1-p)^{N-k} \text{ , since } k << N$$

$$= \frac{T^k\lambda^k(\delta t)^k}{\lfloor k\,(\delta t)^k}\left(1 - \frac{\lambda T}{N}\right)^{N-k}$$

$$= \frac{(\lambda T)^k\,e^{-\lambda T}}{\lfloor k} \text{ in the limit.}$$

We have already made use of this result in § 6.6.

If we introduce the average time T_m between events, so that

$$T_m = \frac{1}{\lambda} \text{ , we get } \frac{1}{\lfloor k}\left(\frac{T}{T_m}\right)^k e^{-\frac{T}{T_m}}.$$

The form of these functions is shown in Fig.12.1.

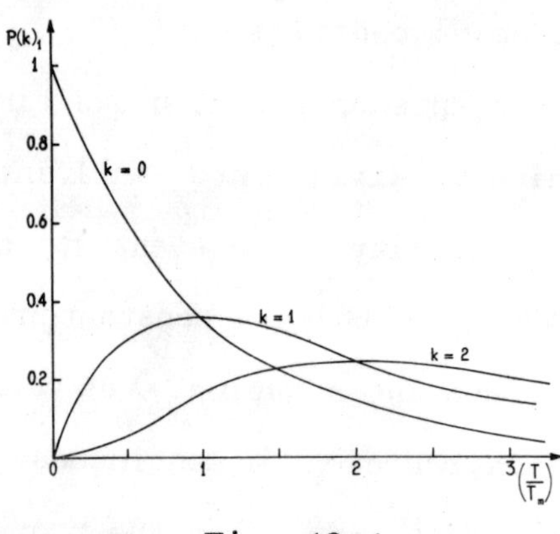

Fig. 12.1

It is important to distinguish between the expectation of k for a given T, $E[k]$, that is the average value of k to be expected, which is of course T/T_m, and the probability of any particular value of k , which is given by the formulae above.

This distribution is discussed in most books on probability theory and is tabulated in Parzen, Ref. 2.1.

12.5. Estimation of parameters from a sample

The problem is this: given a large number of possible events (the population), we select at random a small number (the sample) for examination or measurement, and from this sample we evaulate certain parameters. What conclusions about the parameters of the population can be drawn from the parameters of the sample? In our context, for example, we examine a short time stretch of a random function, and we want to know how far its properties can be taken as representative of the hypothetical function of infinite duration.

We may state two elementary conditions which we would like our estimate to meet:

(1) that the estimated parameter deduced from a single sample should converge to the population parameter as the size of the sample is increased. Such an estimate is said to be 'consistent'.

(2) that the estimated parameter obtained as an average over a number of small samples should converge to the population parameter as the number of samples is increased. Such an estimate is said to be 'unbiased'.
These conditions, though simple and obvious enough, are not al-

ways met by unsophisticated methods of analysis. To illustrate
these terms, we shall consider the case of discrete observations,
and the calculation from them of the mean and the variance.

Let the population Y have a probability density
function $p(y)$, and let us take the mean value $\bar{y}$ to be zero.
Then $E[y] = \bar{y} = \int_{-\infty}^{\infty} y \cdot p(y) dy = 0$.

$$E[y^2] = E[(y - \bar{y})^2] = \int_{-\infty}^{\infty} y^2 p(y) dy = \sigma y^2 .$$

These parameters of the population are to be regarded as fixed,
but not necessarily known, quantities.

Let us now take a sample of size n from this population giving
values $x_i = x_1, x_2$ etc.

Then the sample mean $\bar{x} = \dfrac{\sum x_i}{n}$

and the sample variance $\sigma_x^2 = \dfrac{\sum (x_i - \bar{x})^2}{n}$.

We now relate $\bar{x}$ to $\bar{y}$, and σx to σy .

Each member x_i of the sample, being drawn from the population,
has the properties of a random variable with the parameters of
the population. Thus $E[x_i]$ representing the mean value of a
sample over all possible values, weighted by the probability,
will be $\bar{y}$, that is 0 . Similarly $E[x_i^2]$ for any member of the
sample is σy^2.

Hence $E[\bar{x}] = E\left[\dfrac{\sum x_i}{n}\right] = \dfrac{1}{n}\{E[x_1] + E[x_2] + ...\} = 0 = \bar{y}$.

That is, the expectation of $\bar{x}$ is the value $\bar{y}$, regardless of the sample, and obviously $\bar{x} \rightarrow \bar{y}$ as the size of sample increases. We can also find the variance of the mean, as the mean square over **m** values.

$$\sigma_{\bar{x}}^2 \;=\; \frac{1}{m}\sum(\bar{x} - E[\bar{x}])^2 \;=\; \frac{1}{m}\sum\bar{x}^2$$

Now
$$\bar{x}^2 \;=\; \frac{1}{n^2}\left(\sum x_i\right)^2 \;=\; \frac{1}{n^2}\left[\sum x_i^2 + 2\sum x_i x_k\right]$$

and
$$E[\bar{x}^2] \;=\; \frac{1}{n^2}E\left[\sum x_i^2\right] + \frac{2}{n^2}E\left[\sum x_i x_k\right] \;=\; \frac{1}{n}\sigma y^2,$$

since the expectation of the product terms of type $x_i x_k$ is zero, with x_i and x_k as independent events.

So
$$E[\sigma_{\bar{x}}^2] \;=\; \frac{1}{m}E\left[\sum\bar{x}^2\right] \;=\; \frac{1}{n}\sigma y^2 .$$

Hence the variance of $\sigma_{\bar{x}} \rightarrow 0$ as n increases.

We conclude that $\bar{x}$ is a consistent and unbiased estimare of $\bar{y}$.

Now consider the estimate σ_x^2 of the variance.

$$\sigma_x^2 \;=\; \frac{1}{n}\sum(x_i - \bar{x})^2 \;=\; \frac{1}{n}\sum x_i^2 - \bar{x}^2$$

$$\therefore E[\sigma_x^2] \;=\; \frac{1}{n}E\left[\sum x_i^2\right] - E[\bar{x}^2]$$

$$=\; \sigma y^2 - \frac{1}{n}\sigma y^2$$

$$=\; \frac{n-1}{n}\sigma y^2 .$$

Hence mean value of $\sigma_x^2 \longrightarrow \sigma_y^2$ as n is increased; but the expected value of σ_x^2 for any finite sample size is not σy^2. The estimate is therefore consistent, but biased.

The application of this idea to random vibration measurements will be taken up in Chapter 17.

A general treatment of the theory is given in Cramer Ref. 2.2, and the applications are discussed in Bendat and Piersol, Ref. 1.6.

Chapter 13

APPLICATIONS II : FAILURE DUE TO RANDOM VIBRATION 1

13.1. Response and failure

We are interested in random vibration largely because of its ability to cause failure, and an obvious application of our response analysis is therefore to predict whether, under given conditions, failure will or will not occur. Such application is not as simple as might be expected, but we can certainly make some progress towards it.

Failure can arise from a variety of causes, and it must be remembered that failure of electronic equipment due to excessive acceleration can be just as disastrous as structural failure due to excessive stress. But whatever the mechanism, in considering failure we shall be more concerned with the peak values of a quantity, rather than with the values of the quantity itself. Whether we are considering outright failure due to a single high stress, or fatigue failure due to continuous random vibration it will be the distribution of the extreme values which concern us, and not the distribution of the quantity itself.

Knowledge of the spectral density of a quantity leads immediately to its mean-square value, and so – at least if the distribution is Gaussian – to its distribution with respect to time. We must now consider whether it can also lead to the specification of the peak distribution with respect to time.

13.2. Peak distribution

A peak value of $x(t)$ occurs when $x'(t) = 0$, provided that at the same time $x''(t)$ is negative. So the probability that a peak will occur in a given range of values dx during a given interval dt can be written as

$$(1) \quad \Pr[\text{Peak in } dx\, dt] = \Pr\left[\left.\begin{array}{l} x_1 < x(t) < x_1 + dx \\ x'(t) \;=\; 0 \\ x''(t) \;<\; 0 \end{array}\right\} \text{within time } dt\right].$$

This probability could of course be established by the statistical analysis of a large number of records, but we shall hope to establish it by analysis.

Consider the joint probability density, $p(x,x',x'')$ of the three quantities $x(t), x'(t), x''(t)$, defined such that at any instant of time t_1,

$$(2) \quad p(x,x',x'')dx\,dx'\,dx'' = \Pr\left[\begin{array}{l} x_1 < x(t_1) < x_1 + dx \\ x_1' < x'(t_1) < x_1' + dx' \\ x_1'' < x''(t_1) < x_1'' + dx'' \end{array}\right].$$

The probability defined by (1) can be
related to (2) as follows:

Consider the $x'(t)$, t plot as
shown: the quantity x' becomes zero
in the interval dt only if

$$-x''(t_1)dt > x'(t_1) > 0. \tag{3}$$

The probability of this occurring is that of (2) integrated to
include all negative values of x'' , and for each x'' all the
values of x' covered by $x'(t_1)$ in (3).

$$\text{Thus Pr}\big[\text{Peak in } dxdt\big] \;=\; \text{Pr}\begin{bmatrix} x_1 < x(t_1) < x_1 + dx \\ 0 < x'(t_1) < -x''(t_1)dt \\ -\infty < x''(t_1) < 0 \end{bmatrix}$$

$$= \; dx \int_{-\infty}^{0} \left[\int_{0}^{-x''dt} p(x,x',x'')dx' \right] dx'' \tag{4}$$

$$= \; (\text{putting } x' = 0) - dx\,dt \int_{-\infty}^{0} x''p(x,x',x'')_{x'=0}\, dx'' \;.$$

This gives the <u>probability</u> of a peak in the small
region $dx\,dt$; from it we can obtain the <u>number</u> of peaks expected within any range of x , per unit time, by suitable integration. Thus the expected number of peaks per unit time of magnitude greater than x_A is given by

$$E_1\big[\text{peaks} > x_A\big] \;=\; -\int_{x_A}^{\infty}\left[\int_{-\infty}^{0} x''p(x,x',x'')_{x'=0}\, dx''\right]dx \;. \tag{5}$$

Hence the probability of extending any given critical value x_A in any given time can be inferred, provided that $p(x, x', x'')$ is known. Our interest in the latter quantity derives from the fact that in the special but commonly applicable case of a Gaussian random process $\{x(t)\}$ it can be expressed in terms of the spectral density.

13.3. Peak distribution for Gaussian process

For a Gaussian random process $\{x(t)\}$, it can be shown that

$$
(6) \quad
\begin{aligned}
p(x, x', x'')_{x'=0} &= \frac{1}{\sqrt{8\pi^3 \sigma_{x'}^2 (\sigma_x^2 \sigma_{x''}^2 - \sigma_{x'}^4)}} \times \\
&\quad \exp\left[\frac{-\sigma_{x'}^2 \sigma_{x''}^2 x^2 - 2\sigma_{x'}^4 x x'' - \sigma_x^2 \sigma_{x'}^2 x''^2}{2\sigma_{x'}^2 (\sigma_x^2 \sigma_{x''}^2 - \sigma_{x'}^4)}\right],
\end{aligned}
$$

where $\sigma_x^2 = \langle x^2(t) \rangle$, etc.

Substituting in (4) and carrying out the integration with respect to x'' we obtain

$$
(7a) \quad
\begin{aligned}
\Pr[\text{Peak in } dx\, dt] &= dx\, dt \frac{\sigma_{x'}}{2\sqrt{2}\,\pi \sigma_x^2} \cdot \frac{x}{\sqrt{2}\,\sigma_x} e^{\frac{-x^2}{2\sigma_x^2}} \left[1 + \operatorname{erf}\frac{kx}{\sqrt{2}\,\sigma_x}\right. \\
&\quad + \frac{1}{\sqrt{\pi}} \frac{\sqrt{2}\,\sigma_x}{kx} e^{-\frac{k^2 x^2}{2\sigma_x^2}}\Bigg],
\end{aligned}
$$

where $k^2 = \sigma_{x'}^4/(\sigma_x^2\sigma_{x''}^2 - \sigma_{x'}^4)$, and $\quad \mathrm{erf}\ z = \dfrac{2}{\sqrt{\pi}}\displaystyle\int_0^z e^{-t^2}\,dt$. $\qquad(7b)$

This can be integrated to give the expectation of peaks in any range of x and t .

Equation (7) is greatly simplified if we are interested in peak values of x greatly in excess of σ_x : it then becomes possible to neglect the final term in the square bracket and to treat the erf term as unity. With these approximations we have

$$\Pr\left[\text{Peak in } dx\,dt\right] = dx\,dt\ \frac{\sigma_{x'}}{\sqrt{2}\,\pi\,\sigma_x^2}\cdot\frac{x}{\sqrt{2}\,\sigma_x}\,e^{-x^2/2\sigma_x^2} . \qquad(8)$$

Integration is now easy and expectation per unit time of peaks greater than $n\sigma_x$ is given by

$$
\begin{aligned}
E_1 &= \frac{1}{\pi}\frac{\sigma_{x'}}{\sigma_x}\int_{n\sigma_x}^{\infty}\frac{x}{\sqrt{2}\,\sigma_x}\,e^{-x^2/2\sigma_x^2}\,d\!\left(\frac{x}{\sqrt{2}\,\sigma_x}\right)\\
&= \frac{1}{2\pi}\frac{\sigma_{x'}}{\sigma_x}\,e^{-\frac{n^2}{2}} .
\end{aligned}
\qquad(9)
$$

We need then to know only σ_x and $\sigma_{x'}$, and these can be determined easily if the spectral density is known.

Consider for example a stress $s(t)$ which has a constant spectral density S_1 up to a frequency f_1 as shown

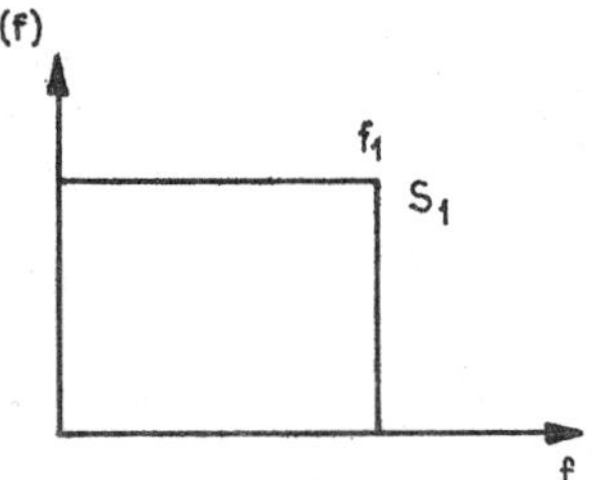

Here

$$E_1 \;=\; \frac{1}{2\pi}\left[\frac{\displaystyle\int_0^\infty 4\pi^2 f^2 S(f)\,df}{\displaystyle\int_0^\infty S(f)\,df}\right]^{\frac{1}{2}} e^{-n^2/2} \;=\; \frac{1}{2\pi}\left[\frac{\displaystyle\int_0^{f_l} 4\pi^2 f^2 S_1\,df}{\displaystyle\int_0^{f_l} S_1\,df}\right]^{\frac{1}{2}} e^{-n^2/2}$$

$$=\; \frac{f_1}{\sqrt{3}}\, e^{-n^2/2}\; .$$

So if $f_1 = 1000\,c/s$ and we wish to know E_1 for $n = 5$,

$$E_1 \;=\; \frac{1000}{\sqrt{3}}\, e^{-12,5} \;=\; 0.00215 \quad \text{peaks per second}$$

If for example $\sigma_s = 5\,tonf/in(\doteqdot 100MN/m^2)$ and the material is assumed to fail when a peak greater than $25\,tonf/in^2$ is reached, then the probability that failure will occur within one minute is approximately one-tenth.

13.4. Peak distribution and structural failure

Structural failure due to random vibration usually takes the form of a fatigue failure, arising from prolonged exposure to the fluctuating stress. Our next requirement therefore is a valid criterion of failure which will enable us to predict the life of a given component when exposed to randomly varying stress of given characteristics. Unfortunately no entirely satisfactory criterion exists, although there have been many attempts to establish one. Some guidance can be obtained from an adaptation of the Miner damage criterion, originally intended to apply to a component subjected to periods of cyclic stressing of different levels of intensity.

The Miner criterion is based on the assumption that if failure under a given stress amplitude occurs after N cycles, then damage occurs uniformly cycle by cycle, so that during one cycle $1/N$ of the total damage occurs. Thus if $N_1, N_2, \ldots$ are the total lives at stress amplitudes $s_1, s_2, \ldots$, then complete failure after $n_1, n_2, \ldots$ cycles at each stress will occur when

$$\sum_r \frac{n_r}{N_r} = 1 \, . \tag{11}$$

This can be adapted to random vibration by letting $s_1, s_2, \ldots$ refer to peak levels rather than to amplitudes. Our peak level analysis then becomes directly relevant.

$$\text{Let } \Pr\left[\text{Peak in } ds\,dt\right] = p_{pk}(s)\,ds\,dt \, . \tag{12}$$

Then the expected number of peaks in the range $s_1 < s < s_1 + ds$ during time T is given by

$$E_T = T p_{pk}(s_\ell)\,ds \, . \tag{13}$$

Now the $s-N$ curve for the material will indicate the total number of cycles N_1 for complete damage under stress s_1 , so the expected damage in time T due to stresses in the range $s_1 < s < s_1 + ds$ is

$$\frac{1}{N_1} T p_{pk}(s_\ell)\,ds \, ;$$

the total damage in time T due to all stress levels is therefore

$$(14) \qquad \int_0^\infty \frac{1}{N(s)} T p_{pk}(s) ds \; ,$$

and for failure this must be unity.

Thus the expected life T_F is given by

$$(15) \qquad T_F = \frac{1}{\displaystyle\int_0^\infty \frac{p_{pk}(s) ds}{N(s)}} \; .$$

Chapter 14

APPLICATIONS III : FAILURE DUE TO RANDOM VIBRATION 2

14.1. Introduction

The preceding chapter on failure has shown how an expected life under random loading can be calculated when there is either a threshold stress at which failure occurs with probability one, or where the Miner criterion of cumulative damage is valid. The present chapter will consider the applicability of the Miner criterion in the light of certain probabilistic models of the fatigue process in polycrystalline materials. Experimental studies and proposed modifications of the Miner criterion will also be discussed.

By way of introduction, it is advisable first to clarify what is meant by "expected life". In the past many distinct physical processes have been collectively known as "fatigue". This has led to great confusion in comparing result and predictions. It is now becoming customary to divide any fatigue process into at least two successive stages. Stage I is known as crack nucleation. It appears to involve the formation of a spatial distribution of small-scale slip bands and their gradual deteriation into extrusions, fissures, or micro-notches. These defects are crystallographically oriented and can be removed from a surface by electropolishing. Stage I is completed by the coalition of these interacting defects into a so-called micro-crack, which is not crystallographically oriented. The point in space and time where this coalition occurs is obviously a statistical question. Stage II involves the propagation of the nucleated crack through the structure of the material. During this stage such macroscopic properties as work hardening, yield stress, notch sensivity and stress gradient begin to play an important part. The behaviour in stage I is unmistakably statistical, whereas deterministic models of stage II can be reasonably successful. Stage II can occupy a considerable fraction of service life in low strength, "ductile" materials. However, stage I becomes more important in uniformly stressed, non-redundant machine components, and also in the case of high-strength aluminium alloys where the rate of crack propagation is some 150 times greater than that in mild steel (ref. 14.1). The present chapter is primarily concerned with stage I, so expected life will be taken to mean the completion of this stage.

14.2. Probabilistic models of stage I

Freudenthal's early paper (ref. 14.2) emphasizing the statistical aspect of fatigue in metals was published prior to the advent of the electron microscope but postulates a model close to the foregoing description of stage I which has evolved from observations using the electron microscope. His model is in fact applicable to stage I, for it is postulated that when

the expected life is attained, at least one of a prescribed number m of elementary events occurs. In the light of the present model of crack nucleation the Freudenthal model can be
reinterpreted to mean that at least one pair of extrusions or
fissures must suddenly coalesce to form the micro-crack. Then
if p_1 is the probability (dependent on stress level) of one
coalescence during one load cycle we have the Freudenthal expression

$$P = 1 - (1 - p_1)^{mn_1}$$

as the probability of reaching the end of stage I after n_1 cycles. Since p_1 is small and mn_1 is large we have approximately

$$P(n_1) = 1 - e^{-mp_1n_1}.$$

Hence the mean value N_1 of n_1 is $(mp_1)^{-1}$. The reliability
function $R(n)$ is defined as $1 - P(n)$, so the above simple model
leads to the standard reliability expression

$$R(n) = \exp(-n/N).$$

The constant hazard rate is $mp_1 = N_1^{-1}$. This model predicts a
standard deviation of lives about the mean equal to the mean
value, whereas in practice this scatter is about 0.4 times the
mean. Nevertheless, the result is of the right order and it is
instructive to consider what the model predicts if a programme

of constant stress amplitudes $s_1, \ldots, s_t$ corresponding to mean lives $N_1, \ldots, N_t$ are applied for $n_1, \ldots, n_t$ cycles, respectively, where $n_i/N_i \ll 1$. (This is known as a t-level block programme). Now we have

$$R(n) = e^{-\frac{n_1}{N_1}} e^{-\frac{n_2}{N_2}} \ldots \doteqdot \left(1 - \frac{n_1}{N_1}\right)\left(1 - \frac{n_2}{N_2}\right)\ldots$$

so to the first order

$$R(n) = 1 - \sum_{i=1}^{t} n_i/N_i \ ,$$

where $n = \sum_{i=1}^{t} n_i$ is the total number of cycles. Hence

$$R \doteqdot 0 \quad \text{when} \quad \sum n_i/N_i = 1 \ ,$$

$$R \doteqdot \frac{1}{2} \quad \text{when} \quad \sum n_i/N_i = \frac{1}{2} \ .$$

The expression $\sum n_i/N_i = 1$ occurs in the original criterion of Miner (or Palmgren), which did not concern itself with probable failure but with certain failure $(R = 0)$.

The above model would lead one to expect a median value of $\sum n_i/N_i = \frac{1}{2}$ in a series of experiments. Since the two original parameters m and p_1 appear as a product in the final expression, there is no freedom to fit observed trends in both mean life and scatter as the stress intensity is varied.

As a second simple (Bernouilli-type) model, let there be a constant probability p_1 (depending on stress intensity) of a microfailure during each cycle and assume that a total of m micro-failures must occur before the end of stage I. Then the probability $p(n)$ of stage I lasting n cycles is

$$p(n) = p_1 \binom{n-1}{m-1} p_1^{m-1} (1 - p_1)^{n-m}$$

where $\binom{r}{s}$ denote the binomial coefficients. Here we have the standard Pascal distribution with mean $N_1 = mq_1/p_1$ and variance $V_1 = S_1^2 = mq_1/p_1^2$, where $q_1 = 1 - p_1$. For the typical experimental values $N_1 = 1.3 \times 10^6$, $S_1 = 0.52 \times 10^6$ we have the estimates $m = 6$, $p_1 = {} = 4.8 \times 10^{-6}$. Now the calculation of the expected life for a multi-level block programme is cumbersome, but the method can be illustrated by the case of two levels in which $n_1 = 3 \times 10^6$ cycles of stress amplitude s_1 with expected life $N_1 = 5 \times 10^6$ and $p_1 = 1 \times 10^{-6}$ are followed by n_2 cycles at a stress amplitude s_2 for which $N_2 = 0.25 \times 10^6$ with $p_2 = 20 \times 10^{-6}$. The probability $p(n_2)$ of failure after these n_2 cycles at level 2 is now

$$p(n_2) = \sum_{r=0}^{m-1} \binom{n_1}{r} p_1^r q_1^{n_1-r} \cdot p_2 \binom{n_2 - m + r}{m - r - 1} p_2^{m-r-1} q_2^{n_2 - n_1 - 1}.$$

The expected life $E(n_2) = \sum_{n_2=1}^{\infty} n_2 p(n_2)$ is given by

$$E(n_2) = \sum_{r=0}^{4} \binom{n_1}{r} p_1^r q_1^{n_1-r} (m-r) \frac{q_2}{p_2},$$

$$= \frac{q_2}{p_2} \left(5b(0) + 4b(1) + 3b(2) + 2b(3) + b(4) \right).$$

The binomial expressions $b(r) = \binom{n_1}{r} p_1^r q_1^{n_1-r}$ can be approximated by the Gaussian probability expression with mean $p_1 n_1 = 3$ and standard deviation $(n_1 p_1 q_1)^{1/2} = \sqrt{3}$. Thus we find $b(0) = 0.07$, $b(1) = 0.12$ etc. $E(n_2) = 2.07\, q_2/p_2 = 0.41\, N_2$, and hence

$$(8) \qquad \frac{n_1}{N_1} + \frac{E(n_2)}{N_2} = 1.01 .$$

At similar calculation with $n_1 = 1 \times 10^6$ yields

$$(9) \qquad \frac{n_1}{N_1} + \frac{E(n_2)}{N_2} = 0.99 .$$

Although these sums agree closely with the Miner criterion, it should be borne in mind that experimental data on the scatter of fatigue lives under a constant stress amplitude indicate that it is unrealistic to assume that m is a constant independent of stress level.

Kyrala (ref. 14.3) has introduced the idea of a stress image point, which under the action of the loading cycles executes a random walk in the stress plane (as the result of accumulating residual stresses). The expected life is expended when the stress image point reaches a boundary circle (absorption barrier). The idea of using the probability function $p(r,n)$ to describe the distance r of the image point from the origin after n cycles has been exploited by Tunguskova (ref. 14.4). A volume V contains kV sub-units, called defects, in which residual stresses may develop during the load cycle. The residual stress in each of the kV/m "defect totalities" is described by the random walk process of ref. 14.3, which yields an expression $p_0(s,n)$ for the probability of each defect totality failing. The probability P of completing stage I, defined as the failure of at least one of the kV/m defect totalities is given by

$$P = 1 - (1 - p_0(s,n))^{kV/m} .$$

With the assumption that at any fixed stress amplitude the step distance s in the random walk is constant, it is shown that

$$p_0(s,n) \fallingdotseq 1 - a\,e^{-bs^2 n} ,$$

i.e. the probability of failure of the sub-units increases exponentially with load cycles. It is finally shown that under a multi-level block programme the first stage will be completed when $\sum n_i/N_i = 1$ (to the first order). It is to be noted that here also the number of defects m in a basic unit is assumed to be independent of stress amplitude, as in the preceding models.

In ref. 14.5 the fatigue is again assumed to consist in passing through a fixed number n of states $E_0, E_1, \ldots \ldots, E_{n-1}$. A more general (homogeneous) Markov model is obtained by introducing the transition probabilities $p_{ij}(t)$ of passing from state E_i to state E_j in any time interval t. Only forward transitions are envisaged. These $\frac{1}{2}n(n+1)$ quantities must satisfy the Kolmogorov forward differential equation and hence represent only n stress-dependent parameters $\lambda_0, \lambda_1, \ldots, \lambda_{r-1}$. At any sinusoidal stress amplitude s_r the mean M_r and variance V_r of the fatigue life are given by

$$(10) \qquad M_r = \sum_{i=0}^{n-1} \frac{1}{\lambda_{ir}} \; , \qquad V_r = \sum_{i=0}^{n-1} \frac{1}{\lambda_{ir}^2} \; .$$

This embarrassingly large number of parameters is reduced to two by assuming that

$$(11) \qquad \lambda_{ir} = \lambda_{or}(1 + i)^{q_r} \; .$$

The successive states under a four-level block programme is then computer simulated using the Monte Carlo technique. The results are compared with actual experimental tests on an alloy at 700° C. Using the published values of the median lives N_i, one finds that the model predicts 50% probable failure when $\sum n_i / N_i = 1.06$

14.3. Experimental aspects

In spite of a very large volume of test results in which the values of $\sum n_i / N_i$ were close to one at failure, warnings were voiced as early as 1954 (ref. 14.6) that when the loading contains many stress peaks below the fatigue limit a so-called stress interaction effect could lead to failure when $\sum n_i / N_i \ll 1$. Gassner was concerned with estimating the life of an automobile front-wheel suspension system in which the vast majority of loading peaks are below the fatigue limit and are not damaging according the Miner criterion. His solution was purely empirical and consisted in simulating the service stresses by an 8-level block programme. Several laboratories have studied the effects of changing the order of the levels in the blocks. Some authorities claim that the order is not critical provided the total life consists of at least 20 blocks. Others are more cautious and claim that the positions of the levels in each block should be randomised, whereas Schijve (ref. 14.7) claims that procedures for establishing a programmed load sequence to agree with the results of random loading have not yet been ascertained (1963).

If a direct study of damage due to random fatigue loading is to be made, the first step consists in deciding what are the relevant parameters. Expression 13.3 (7) indicates that σ_x, σ_x', σ_x'' are sufficient parameters for describing the peak distribution in a Gaussian process. Vibration theory has shown us that most mechanical responses will be modal with a characteristic frequency and bandwidth. With the aim of replacing the above quantities σ_x' and σ_x'' by a bandwidth parameter, it is expedient to rewrite 13.3 (7) in terms of the dimensionless variable $\eta = x/\sigma_x$. Now the probability of a peak between η and $\eta + d\eta$ is $p(\eta)\, d\eta$ where

$$p(\eta) = \frac{1}{\sqrt{2\pi}} \left[\varepsilon e^{-\frac{\eta^2}{2\varepsilon^2}} + \sqrt{1+\varepsilon^2}\, \eta e^{-\frac{\eta^2}{2}} \int_{-\infty}^{+\eta(1-\varepsilon^2)^{1/2}/\varepsilon} e^{-t^2/2}\, dt \right] ,$$

i.e. in addition to the scaling parameter $\sigma_x = <x^2>^{\frac{1}{2}}$, the peak distribution contains only one parameter

$$(12) \qquad \varepsilon^2 = \frac{\sigma_x^2 \sigma_x''^2 - \sigma_x'^4}{\sigma_x^2 \sigma_x''^2} \, .$$

Recalling from Chapter 9 that

$$(13) \qquad \sigma_x'^2 = \int_0^\infty 4\pi^2 f^2 S_x(f) df \, , \qquad \sigma_x''^2 = \int_0^\infty 16\pi^4 f^4 S_x(f) df \, ,$$

we see that ε characterises the shape of the power spectrum $S_x(f)$. It is clear that $0 \leqslant \varepsilon \leqslant 1$. For $\varepsilon = 0$ we have the Rayleigh distribution

$$(14) \qquad p(\eta) = \eta\, e^{-\eta^2/2} \, ,$$

and for $\varepsilon = 1$ we have the Gaussian distribution

$$(15) \qquad p(\eta) = \frac{1}{\sqrt{2\pi}}\, e^{-\eta^2/2} \, .$$

There are undoubtely situations in which it is more efficient to estimate ε from the shape of the experimentally measured curve $p(\eta)$, rather than use the experimental value of $S_x(f)$, because the high frequency tail on the latter can affect σ_x'' critically.

     A further interpretation of ε is gained by considering the fraction r of the maxima that are of negative sign. For a narrow-band process that looks like a sine wave with modulated amplitude we expect $r = 0$. An increase in bandwidth leads to an increase in r. It can be shown using 13.4 (7) that

$$r = \frac{1}{2}\left[1 - \frac{\sigma_x'^2}{(\sigma_x \sigma_x'')}\right], \qquad (16)$$

and then comparison of (16) and (12) yields

$$r = \frac{1}{2}\left[1 - (1 - \varepsilon^2)\right]^{1/2},$$

or

$$\varepsilon^2 = 1 - (1 - 2r)^2.$$

For obvious reasons r is called the regularity factor.

As an example of how ε is related to the bandwidth consider the narrowband process resulting from passing white noise through an ideal bandpass filter of centre frequency f_0 and bandwidth B. Direct evaluation of (12) using either of the methods in Chapter 9 yields

$$\varepsilon \doteqdot \frac{1}{\sqrt{3}}\frac{B}{f_0}.$$

$$\left(\text{Actually}\quad \varepsilon^2 = \left(\frac{1}{3}\delta^2 + \frac{1}{60}\delta^4\right)\bigg/\left(1 + \frac{1}{2}\delta^2 + \frac{1}{80}\delta^4\right)\ \text{where}\ \delta = B/f_0\right).$$

If a one-freedom system were used instead of the ideal filter, we would use the effective noise bandwidth equal to $(\pi/2)$ times the half-power bandwidth defined in Chapter 4. Hence

$$\varepsilon \doteqdot 0.85/Q$$

where Q is the quality factor.

The experimental work at Glasgow University on narrowed random fatigue is now centred about the effects of σ_x and ε on the values of $\sum n_i/N_i$ at failure.

Bibliography for Chapter 14

14.1   Frost N.E. and Marsh K.J.: Design to Prevent Fatigue
       Failure, in Fracture.
       Second Tewkesbury Symposium, Butteworth, London
       1969.

14.2   Freudenthal A.M.: Statistical Aspects of Fatigue of
       Materials.
       Proc. Royal Soc. London, Vol. A 187, pp. 416-429,
       1946.

14.3   Kyrala A.: Grundlagen einer stochastischen Dauerfestig-
       keitstheorie.
       Osterreich. Ing.Arch. Vol. 14, No.3, pp. 204-218,
       1960.

14.4   Tunguskova Z.G.: A Stochastic Approach to the Analysis
       of Fatigue Failure in Metals (in Russian).
       Vestnik Moskovskogo Universiteta, Ser. 1, No.4,
       pp. 86-90, 1968.

14.5   Kogaev V.P.: Simulation of Metal Fatigue by the Monte
       Carlo Method.
       Industrial Laboratory, Vol. 34, No. 7, pp.990-994,
       1968.

14.6   Gassner E.: Betriebsfestigkeit. Eine Bemessungsgrund-
       lage für Konstruktionsteile mit statisch wechseln-
       den Betriebsanspruchung, Konstruktion, Vol. 6,
       Heft, pp. 97-104, 1954.

14.7   Schijve J.: Fatigue Life and Crack Propagation under
       Random and Programmed Load Sequences, Current
       Aeronautical Fatigue Problems.
       Pergamon Press, 1965.

Chapter 15

APPLICATIONS IV : SIMULATION

15.1. The problem

Many of the engineering systems which suffer from random vibration are of considerable complexity, so that fully instrumented testing in service is just not practicable. Also sub-components of such complex systems are often separately supplied, so that there is a need to demonstrate their capacity for survival by tests which take place outside the parent equipment. Such considerations have led to an interest in simulating in the laboratory vibration conditions which occur in service. It is therefore important to consider how vibration conditions can be simulated, and to consider in particular how this can be done economically, using relatively simple testing equipment.

Ideally one would wish to apply in the laboratory the forces which would be experienced in service – or at least forces which have at all frequencies the same spectral densities and cross spectral densities. As however the forces may arise partly from fluctuating air pressures and partly from combustion processes elsewhere in the system these may well be impossible to isolate and instrument. It is therefore likely to be a more reasonable objective to reproduce the correct response rather than the correct excitation, and this is the problem that we shall consider here.

It will simplify our exposition of the problem if the component considered is a simple one: we shall consider the problem of simulating the observed motions of a simple beam-like system.

15.2. Response in terms of normal modes

Analysis of this problem will benefit greatly by treatment in terms of the normal modes of the system. We shall first establish some basic results for the response of a simple beam to random excitation.

Consider a beam of length ℓ ; the displacement $w(x,t)$ at any point x at time t can be expanded in the form

$$w(x,t) = \sum_r \xi_r(t) w_r(x), \tag{1}$$

where $w_r(x)$ gives the shape of the rth normal mode, and the $\xi_r(t)$ are normal coordinates indicating the contribution of each mode at time t .

The cross-correlation function of the displacements at any two points x_i , x_j is given by

$$R_{w_i w_j} = \left\langle \sum_r \xi_r(t) w_r(x_i) \sum_s \xi_s(t+\tau) w_s(x_j) \right\rangle$$

$$= \sum_{r,s} w_r(x_i) w_s(x_j) R_{\xi_r \xi_s} . \tag{2}$$

Transformation of (2) gives the corresponding result relating the spectral densities of the actual displacements of those of the normal coordinates

$$(3) \qquad S_{w_i w_j} = \sum_{r,s} w_r(x_i) w_s(x_j) S_{\xi_r \xi_s} .$$

It may be noted that cross-correlations between normal coordinates contribute even towards the direct spectral density of the response at any point.

The generalised forces Z_r corresponding to the ξ_r must be such that

$$\sum_r Z_r \delta \xi_r = \sum_m P_m \delta w(x_m) ,$$

which, by (1), implies

$$Z_r = \sum_m P_m w_r(x_m) ,$$

so that

$$(4) \qquad S_{Z_r Z_s} = \sum_{m,n} w_r(x_m) w_s(x_n) S_{P_m P_n} .$$

The response in the rth mode must satisfy

$$(5) \qquad \ddot{\xi}_r + 2\zeta_r \omega_r \dot{\xi}_r + \omega_r^2 = Z_r/M_r ,$$

when ω_r is the rth natural frequency, ζ_r is the damping ratio (damping being assumed viscous and without coupling between modes), and M_r is the generalised mass). From (5) the receptance α_r connecting ξ_r and Z_r is given by

$$\alpha_r = \frac{1}{M_r(\omega_r^2 - \omega^2 + 2i\zeta_r\omega_r\omega)} \, . \tag{6}$$

This simple notation is made possible by the fact that there is no coupling between the rth and sth modes, and the same fact also gives a simple form to the spectral density relation

$$S_{\xi_r\xi_s} = \alpha_r^*\alpha_s S_{z_r z_s} \, . \tag{7}$$

The simplicity of (7) – the absence of a summation sign – is in fact the reason for the modal approach to response. Although the complete response relationship requires equations (3), (4), and (7), the dependence on frequency is confined to the single equation (7), where moreover the α's have a particularly simple form.

15.3. Simulation

Suppose that a given beam A in its service environment is subjected to a set of forces $P_m^A(t), m=1,2,...,$ which are unknown, and that displacements $w^A(x,t)$ occur at the various points x . We require to reproduce the same motions in an identical beam B by applying suitable forces. We shall regard the simulation as exact if for all pairs of points x_i, x_j and for all frequencies ω , the spectral densities in the two cases are identical, i.e. if

$$S^B_{w_i w_j}(\omega) = S^A_{w_i w_j}(\omega) \, . \tag{8}$$

To satisfy (8) requires the matching of the spectral densities of the displacements $w(x,t)$ at an infinite number of points along the beam: it can be seen from (3) that the satisfaction of (8) in respect of an infinite number of points requires the satisfaction of a similar relationship between the spectral densities of an infinite number of normal coordinates

$$(9) \qquad\qquad S^{B}_{\xi_r \xi_s}(\omega) = S^{A}_{\xi_r \xi_s}(\omega) .$$

This, by (7) can only be achieved by matching an infinite number of generalised forces, satisfying

$$(10) \qquad\qquad S^{B}_{z_r z_s}(\omega) = S^{A}_{z_r z_s}(\omega) .$$

And this can only be satisfied, by (4), if

$$(11) \qquad\qquad S^{B}_{P_m P_n}(\omega) = S^{A}_{P_m P_n}(\omega)$$

i.e. if all applied force spectral densities are matched.

Thus the only way to simulate exactly the motions experienced in service is to apply exactly the forces occurring in service. It is not for example possible to reproduce the effects of a complicated system of forces exactly by means of a single applied force.

As force systems will usually be complex and as the application and control of a great number of forces during a test is not practicable we must evidently consider the possibility of approximate simulation.

15.4. Approximate simulation

One obvious method of approximation is to require the matching of spectral densities of displacements at only a finite number of points n. But clearly n would need to be very large to give realistic simulation, and in general such a test would require the control of n^2 spectral densities of n separate forces.

If however we aim to simulate the spectral densities of n normal coordinates a very much smaller number will usually suffice to give good simulation. Consider equation (7): each α_r will have the form of a typical single-freedom resonance curve, having a large magnitude only at or near the corresponding natural frequency ω_r. The product $\alpha_r^* \alpha_s$ can be large only if (a) $r = s$ or (b) if ω_r and ω_s are very close together. Thus in many

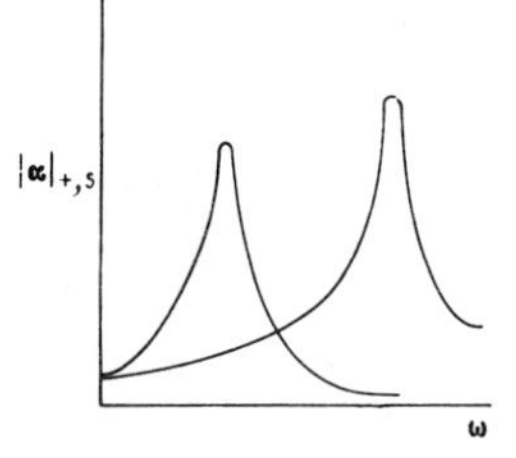

practical circumstances only a few ξ_r will be of any magnitude at any given frequency, and so only a few modes will contribute appreciably to the summation for $S_{w_i w_j}$ in (3). In fact very often a single mode will predominate.

Text excitations which aim to simulate correctly only one or two normal modes are therefore of considerable practical interest.

(a) <u>Single Mode Simulation</u>

Where damping is small and peaks well separated, so that a single mode predominates near any natural frequency (3) can be replaced by

$$(12) \qquad S_{w_i w_j}(\omega) = w_r(x_i) w_r(x_j) S_{\xi_r}(\omega) .$$

This will of course give inaccurate values for spectral density away from resonance, but such values will always be small and will have little effect on the total response. The values of r taken in (12) at any ω will always be that of the nearest ω_r .

If (12) holds it is certainly possible to impose a given $S_{w_i w_j}(\omega)$ spectrum by means of a single applied force. By (12) to impose the required $S_{w_i w_j}(\omega)$ at any frequency requires only that $S_{\xi_r}(\omega)$ has the correct value at that frequency; this can be assured, as (7) shows, by ensuring that $S_{z_r}(\omega)$ has the correct value at that frequency; this can be assured, as (4) shows, using a single applied force $P(t)$, at any point, provided that $S_P(\omega)$ has the correct value at that frequency.

It is easy enough to adjust the force spectrum in this way. If the direct spectral density $S_{w_i}(\omega)$ at any single point x_i is correctly adjusted then the modal spectral density $S_{\xi_r}(\omega)$ must be correct. If $S_{\xi_r}(\omega)$ is correct then all other $S_{w_i w_j}(\omega)$ must be correct also. So if the force spectrum is such as to give the correct response spectral density

at any arbitrarily chosen measurement point, then the whole motion - in this simple case - will be correctly simulated.

(b) <u>Two Mode Simulation</u>

A similar technique based on the assumption that two modes contribute to the response at all frequencies can be developed in
a similar way.

For two modes (3) becomes

$$S_{w_i w_j} = w_r(x_i) w_r(x_j) S_{\xi_r} + w_r(x_i) w_s(x_j) S_{\xi_r \xi_s}$$

$$+ w_s(x_i) w_r(x_j) S_{\xi_s \xi_r} + w_s(x_i) w_s(x_j) S_{\xi_s} .$$

Thus to impose the four correct spectral densities in respect of
all pairs of points x_i, x_j the four modal spectral densities S_{ξ_r},

$S_{\xi_r \xi_s}$, $S_{\xi_s \xi_r}$, S_{ξ_s} must be correctly adjusted. This, by (7) re-

quires that the corresponding spectral densities for z_r, z_s be
correctly adjusted, and this can be done, by (4), by ensuring
that the four spectral densities of any pair of applied forces
P_m , P_n are correct.

Again <u>all</u> $S_{w_i w_j}$ will be correct if the four modal spectral densities are correct: these will be correct if the
four displacement spectral densities of <u>any two</u> arbitrarily chosen measuring points are correctly adjusted.

15.5. Practical considerations

The above analysis has indicated that correct reproduction of n normal modes should be possible if n applied forces are used – though suitable control of n^2 spectral densities may not always be easy. But there are a number of other difficulties which may arise in practice and should be mentioned here.

For example, a mode cannot easily be excited by a force which acts near a node, or normal to the direction of displacement. It may therefore be necessary to have alternative excitation points. And we have assumed here that systems A and B are absolutely identical. In practice this will not be the case and it may not be either easy or useful to compel B to have a large $S_{\xi_r}(\omega)$ at ω_r^A if ω_r^A and ω_r^B differ to any extent. Our analysis is to be thought of as forming the basis for a test rather than a prescription for the test itself.

But there is a further difficulty. At high frequencies resonances are relatively more closely spaced. Consider a pinned-pinned beam for simplicity: as $\omega_r = r^2 \omega_1$, the ratio

$$\frac{\omega_{r+1} - \omega_r}{\omega_r} = \frac{(r+1)^2 - r^2}{r^2} = \frac{2r+1}{r^2}$$

and so decreases with r. Thus at very high frequencies many resonances crowd into small band widths, and it becomes quite impossible to separate the contributions of the separate modes.

The remedy here is to give up any hope of correct simulation of the separate modes and to require instead only that the mean-square contribution of response over a finite band-width is correctly simulated. Such a technique has been tried successfully during recent years, particularly by members of the Bolt Beranek organisation, for response analysis as well as in simulation.

Bibliography for Chapter 15

Analysis of the simulation problem in terms of normal modes is covered in:

15.1   Robson J.D. and Roberts J.W.: A Theoretical Basis for Simulation of Random Vibrations.
J.M.E.S. Vol. 7, pp. 246-251, 1965.

Some consideration of the effects of modal separation on response is given in:

15.2   Robson J.D.: The Random Vibration Response of a System Having Many Degrees of Freedom.
Aero. Quart. Vol. 17, pp. 21-30, 1966.

Some account of finite-bandwidth response analysis, and examples of its use, may be found in:

15.3   Dyer I.: Response of Space Vehicle Structures to Rocket Engine Noise, Chapter 7, in Crandall (ed) Random Vibration Vol. 2, MIT Press, 1963.

and in:

15.4   Lyon R.H.: Random Noise and Vibration in Space Vehicles.
Monograph SVM1, US Dept. of Defence, 1967.

Chapter 16

APPLICATION V – COMPUTATION EVALUATION OF FREQUENCY SPECTRA BY DIGITAL METHODS

16.1. Introduction

The aim of this chapter is to present a simplified account of the two most common methods of digital spectral analysis. It is of necessity short and omits a great deal of important detail, but a comprehensive bibliography is given which will cover all aspects of the computational techniques.

We have come across two definitions of the "power" spectral density function

$$1: - \qquad S(f) = \lim_{T \to \infty} \left\{ \frac{2}{T} \left| A_T(if)^2 \right| \right\} \qquad (1)$$

where
$$A(if) = \int_{-\infty}^{\infty} x(t)\, e^{-i2\pi ft}\, dt \qquad (2)$$

and $A_T(if)$ is the Fourier transform of the function $x_T(t)$ which is equivalent to $x(t)$ over the range $-T/2 < t < T/2$ and zero elsewhere. The substitution of $A_T(if)$ is necessary, otherwise for $x(t)$ stationary, equation (2) will not converge.

$$2: - \qquad S(f) = 2 \int_{-\infty}^{\infty} R(\tau)\, e^{-i2\pi ft}\, d\tau \qquad (3)$$

where $R(\tau)$ is the correlation function.

The spectrum defined by equations (1) and (3) is the "one-sided" spectrum for positive frequencies only.

The two methods of spectral analysis which we will discuss are based on the above definitions. Historically a method based on equation (1) came first, but because of the number of computational operations involved in the calculation of the Fourier transform it was soon superseded by a technique based on equation (3). This method has been extensively used and was written up in a paper by Blackman and Tukey [1] published in 1958.

Up until 1965 this was the only method which could be easily handled by a digital computer. However, in 1965, Cooley and Tukey [2] published an Algorithm for the evaluation of the discrete Fourier transform and it now became possible both in speed and economy to evaluate the spectrum by the direct method of equation (1).

16.2. Discrete data

The definitions for the correlation function, $R(\tau)$, and the frequency spectrum $S(f)$, have been based on a continuous (in) finite signal with the Fourier transform of the correlation function (3) defined for all lag values τ .

Now we must introduce the idea of discrete, finite data, i.e. we must be prepared to carry out analysis work on a series of discrete sampled values of a signal of finite length.

That is, we have a **signal** $x(t)$ in the interval $0 \leq t \leq T$ and we sample values $x(t)$ at equidistant intervals Δt to end up with a sequence of N discrete observations $x_1, x_2, \ldots x_N$. Before proceeding further it is necessary to state without proof some theorems and ideas related to the handling of discrete data.

The sampling theorem attributed to Shannon [3] states that if the power in $x(t)$ is limited to a band less than $\frac{1}{2\Delta t}$ **Hz** then sampling at an interval of Δt enables one to reconstruct the signal uniquely.

Secondly, if $A(if)$ and $x(t)$ are an integral Fourier transform pair

$$A(if) = \int_{-\infty}^{\infty} x(t)\, e^{-i2\pi ft}\, dt \left.\vphantom{\int_{-\infty}^{\infty}} \right\} \qquad \mathbf{(4)}$$

$$x(t) = \int_{-\infty}^{\infty} A(if)\, e^{i2\pi ft}\, df$$

and we define a discrete Fourier transform to be

$$A(r) \sum_{k=0}^{N-1} x(k)\, e^{-i2\pi kr/N} \qquad (5)$$

$$r = 0, 1, 2, \ldots, N,$$

then provided X_k is sampled from $x(t)$ at **equally spaced** points (Nyquist **samples**) then $A(r)$ is closely related to $A(if)$. It is beyond the **scope** of this paper to expand on this **point** and further information can be obtained from Cochran et al. [4] , 1967 and Cooley, **Lewis** and Welch 1969 [5] .

Lastly, all the **theorems** relating **to the integral**
transform **e.g., convolution,** are applicable to the discrete case.

16.3. Standard method or Blackman — Tukey method of spectral analysis

The correlation function is evaluated from the
sampled data **and the** spectrum through its Fourier transform.
Since the correlation function is an even function of τ , we can
modify equation (3) and define the "one-sided" spectrum to be

$$(6) \qquad S(f) = 4 \int_{0}^{\infty} R(\tau) \cos 2\pi f \tau \, d\tau$$

$$(7) \qquad R(\tau) = \lim_{T \to \infty} \frac{1}{T} \int_{-T/2}^{T/2} x(t) x(t+\tau) dt \; .$$

The continuous signal, $x(t)$ with zero mean value,
is sampled at equal intervals, Δt , to obtain the finite series
x_k for $k = 0,1,2,...,N-1$. The choice of Δt must be such that
the largest frequency in the record

$$(8) \qquad f_c = \frac{1}{2\Delta t} \, Hz \; .$$

It is obvious that whereas the correlation func-
tion defined by (7) is evaluated for all lag values, τ , in the
discrete case of N samples, it will not be possible to obtain
an infinite number of lag values, and moreover, it is usually

not desirable to use lag values longer than about 5% – 10% of the record length.

We therefore replace equation (7) by an apparent correlation function for the discrete case. One common estimator of this function is given by

$$\tilde{R}(r) = \frac{1}{N-r} \sum_{k=0}^{N-r} X_k X_{k+r} \tag{9}$$

for $r = 0,1,2,\ldots,m$, where $\tau = r\Delta t$.

The number of lag values, m, decides the equivalent frequency bandwidth of the spectrum in the frequency interval $(0,f_c)$

$$B_e = \frac{2f_c}{m} = \frac{1}{m\Delta t} \tag{10}$$

and for small statistical uncertainty in the spectrum estimates we should choose $m \ll N$ since the maximum number of degrees of freedom is $2N/m$. But, unfortunately, high resolution will result when m is large, and a compromise situation has developed.

Because the apparent correlation function (9) has no value for $\tau > \tau_{max}$

$$\text{where} \quad \tau_{max} = m\Delta t \tag{11}$$

we must replace this function by a modified correlation function $\hat{R}(\tau)$ which is defined for all τ. This is achieved by multiplying the apparent correlation function $\tilde{R}(\tau)$ by a prescribed even function of τ, $D(\tau)$ subject to the restriction that

(12)
$$\begin{cases} D(0) = 1 \\ D(\tau) = 0 \qquad \tau > \tau_{max} \end{cases}$$

(13)   i.e. $\hat{R}(\tau) = \tilde{R}(\tau) \cdot D(\tau)$.

This new function may not be a good estimate of the true correlation function $R(\tau)$, but its fourier transform gives a very respectable estimate of the true spectrum $S(f)$ [1] .

A typical function for $D(\tau)$, is the lag window, whose use is called "Hamming"

(14)
$$\begin{aligned} D(\tau) &= 0.54 + 0.46 \cos \frac{\pi \tau}{\tau_{max}} & \quad |\tau| &< \tau_{max} \\ &= 0 & \quad |\tau| &> \tau_{max} \end{aligned}$$

Further functions for $D(\tau)$ may be found in references [1] and [6] .

However one of the advantages of the digital technique is that instead of modifying the apparent correlation function as in equation (13) it is possible to evaluate a "rough" spectrum, $\tilde{S}(f)$, from the apparent correlation function and then linearly "smooth" or convolve this spectrum with the spectral window $Q(f)$ corresponding to $D(\tau)$. (The transform of a product of two signals yields the convolution of teh transform of each signal). Moreover, if the rough spectrum is evaluated for a frequency interval

$$\frac{1}{2\tau_{max}} = \frac{1}{2m\Delta t}\ Hz \tag{15}$$

this provides for $m/2$ independent spectral estimates (since those at points less than $\frac{2f_c}{m} = B_e$, apart will be correlated)

and the spectral window simplifies to smoothing adjacent points with the weights

$$0.23 \qquad 0.54 \qquad 0.23 \ .$$

The lag window $D(\tau)$ and the corresponding spectral window $Q(f)$ are shown in figure (1)

$$D(\tau) + 0.54 + 0.46\cos\frac{\pi\tau}{\tau_{max}} \qquad \text{for } |\tau| < \tau_{max}$$
$$= 0 \qquad \text{for } |\tau| < \tau_{max}$$

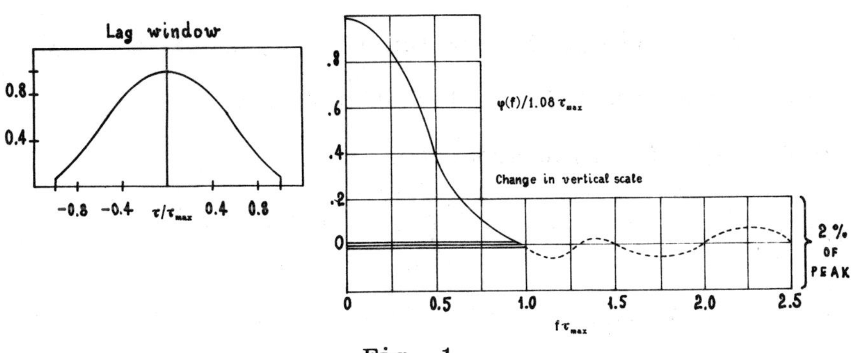

Fig. 1

We can now evaluate the "rough" spectrum by means of the discrete fourier transform which for the case of a real even function can be written as

$$(16) \qquad S(f_r) = 2\Delta t \left[\tilde{R}(0) + 2 \sum_{k=1}^{m-1} \tilde{R}(k)\cos\beta + (-1)^r R(m) \right]$$

where $\quad \beta = 2\pi\tau f_r = 2\pi(k\Delta t)\dfrac{r}{2m\Delta t}$

i.e. $\quad \beta = \dfrac{\pi k r}{m}$

and $\quad f_r = \dfrac{r}{2m\Delta t} \qquad r = 0,1,2,\ldots,m\,.$

The final "smoothed" estimates of the spectrum are obtained by

$$(17) \qquad \hat{S}(f_r) = 0.23\,\tilde{S}(f_{r-1}) + 0.54\,\tilde{S}(f_r) + 0.23\,\tilde{S}(f_{r+1})$$

and $\hat{S}(f_r)$ is our sampled estimate of the true spectrum $S(f)$.

16.4. Direct method using the fast Fourier transform of Cooley and Tukey

If x_k is our sampled signal of N points we can evaluate using the F.F.T. algorithm the discrete fourier transform for this N point sequence.

$$(18) \qquad Ar = \sum_{k=0}^{N-1} X_k e^{-i2\pi k r/N}$$

for $\quad r = 0,1,2,\ldots,N-1\,.$

This seems at first sight to represent no problem, but whereas in the case of the correlation function we were transforming at most 10% of the data, here we transform the complete data. This was the stumbling block in the direct derivation of the power spectrum. The solution of (18) requires us to compute N^2 complex multiply-adds and if N is large, which it usually is, this operation soon became beyond the scope of digital computers both in storage capacity and elapsed time.

The Cooley-Tukey Algorithm for the evaluation of equation (18) cuts this number of operations to $2nN$ provided that the original series consisted of $N = 2^n$ samples. This condition can always be met, either by sacrificing data, or by adding sufficient zeros to complete the sequence.

The algorithm for 2^n data points is described in Appendix 1.

Once we have performed the discrete fourier transform we form the modulus squared of the complex A_r coefficients and evaluate the periodogram $I(f)$ which is not, unfortunately a good estimate of the spectrum.

$$I(f_r) = \frac{\Delta t}{N} \left| A_r \right|^2 \tag{19}$$

$$r = 0, 1, \ldots N/2$$

where $\quad f_r = \dfrac{r}{N \Delta t}$

It is worth noting that although N real data points were transformed to $N-2$ complex frequency values (A_0 and A_n are real), due to the property that $A_r = A_{-r}^* = A_{N-r}^*$ (i.e. sequence X_k is real) and since our maximum frequency in the record in $f_c = \dfrac{1}{2\Delta t}$ values of A_r and $I(f_r)$ are redundant for $\dfrac{N}{2} < r < N$.

Equation (19) will have to be modified by a factor

of 2 to allow for the fact that we wish to evaluate a "one-sided" spectrum.

It remains now a matter of smoothing the spectrum with an appropriate window $\underline{or}$ multiplying the original data by a prescribed function defined for all values of time t . Examples of smoothing functions may be found in reference (7).

APPENDIX 1

The Fast Fourier Transform Algorithm

The version of the algorithm described is for N complex data points, X_k, where $N = 2^n$.

We wish to evaluate the discrete Fourier transform

$$A_r = \sum_{k=0}^{N=1} X_k e^{-i2\pi rk/N} \tag{1}$$

for $r = 0, 1, 2, \ldots N-1$.

Let the series be divided into two series Y_k and Z_k each with $N/2$ points such that

$$\left. \begin{aligned} Y_k &= X_{2k} \\ Z_k &= X_{2k+1} \end{aligned} \right\} \tag{2}$$

$k = 0, 1, 2, \ldots \frac{N}{1} - 1$.

Since Y_k and Z_k are sequences of $N/2$ points each they have discrete Fourier transform defined by

$$\left. \begin{aligned} B_r &= \sum_{k=0}^{\frac{N}{2}-1} Y_k e^{-i4\pi rk/N} \\ \\ C_r &= \sum_{k=0}^{\frac{N}{2}-1} Z_k e^{-i4\pi rk/N} \end{aligned} \right\} \tag{3}$$

$$r = 0, 1, 2, \ldots \frac{N}{2} - 1$$

we require A_r , which we can write in terms of the odd and even numbered points

$$(4) \qquad A_r = \sum_{k=0}^{\frac{N}{2}-1} \left(Y_k e^{-i4\pi rk/N} + Z_k e^{-i2\pi r(2k+1)/N} \right)$$

$$r = 0, 1, 2, \ldots \frac{N}{2} - 1$$

which may be written as

$$(5) \qquad A_r = B_r + C_r e^{-i2\pi r/N} \qquad 0 \leqslant r < N/2 \ .$$

For values of r greater than $N/2$ the discrete fourier transform B_r and C_r repeat periodically, and it follows that

$$(6) \qquad A_{r+N/2} = B_r - C_r e^{-i2\pi r/N} \qquad 0 \leqslant r < N/2 \ .$$

Thus we require to evaluate 2 transforms involving $2\left[\frac{N}{2}\right]^2$ operations and can evaluate A_r with an additional N operations.

Similarly, we can further reduce B_r and C_r and if $N = 2^n$ we can make n such reductions, resulting in a maximum number of $2nN$ operations.

For further details see reference (4).

Bibliography for Chapter 16

[1]   Blackman R.B. and Tukney J.W.: The Measurement of **Power**
Spectra from the Point of View of Communication
Engineering.
Bell System Tech. Journal, March 1958.

(Available in Dover Publications).

[2]   Cooley J.W. and Tukey J.W.: An Algorithm for the Machine
Calculation of Complex Fourier series.
Mathematics of Computation, Vol. 19, pp. 297–301,
April 1965.

[3]   Shannon C.E.: Communications in the Presence of Noise.
Proceedings of the I.R.E., January 1949.

[4]   Cochran W.T. et al.: What is the Fast Fourier Transform?
I.E.E.E. Trans. Au. – No. 2, June 1967.

[5]   Cooley J.W., Lewis P.A.W., Welch P.D.: The Fast Fourier
Transform Algorithm: Programming and Accuracy Considerations in the Calculation of Sine, Cosine,
and Laplace Transforms.
Lect. series on Applications and Methods of Random Data Analysis, Southampton, July 1969.

[6]   Priestley M.B.: Basic Considerations in the Estimation
of Spectra.
Technometrics, Vol. 4 No.4, November 1962.

[7]   Cooley J.W., Lewis P.A.W. and Welch P.D.: The Application of the Fast Fourier Transform Algorithm to
the Estimation of Spectra and Cross Spectra.
Lect. ser. on Applications and Methods of Random
Data Analysis, Southampton, July 1969.

Chapter 17

APPLICATIONS VI : ANALYSIS OF RANDOM DATA

17.1. Introduction

Any attempt to enumerate the reasons for collecting data from random processes would necessarily be incomplete. Often conclusions can be drawn without any involved analysis; an experienced motorist can usually tell by "aural spectrum analysis" when the tappet clearance in his engine has increased just a few thousandth of an inch. Although a trained human operator has been a successful way of keeping complex sets of machinery under surveillance, we are now faced with the problems of fully automated engine rooms (for the understandable reason that trained operators are expensive items). Data on oil flows, steam pressures, bearing vibrations etc. have to be used to make decisions automatically. In some cases the penalty for a wrong decision is quite high, and then it is important to be sure as possible what valid conclusions can be drawn from the data. At this point the statistical nature of data analysis becomes evident. The reduced data based on an observation over a finite time T , i.e. an experiment, contains a component that could vary from sample to sample of length T , the remainder being what one really wants to know. For this reason it has been said that the major function of statistical procedures is to show us what we do not know from

an experiment, not to tell us what we do know. This is far from
being an admission of sterility, it serves as a warning about
the misuse of random data.

The point I will be trying to make in the present
chapter is that random data analysis as distinct from its servant
– data processing – is intimately related to statistical analysis
and in particular that branch known as theory of estimation. There
are of course situations where a relatively crude descriptive
result suffices. Nevertheless, it is well to bear in mind the
statistical nature of any result of processing random data.

The various descriptions or parameters of a ran-
dom process given in the preceding lectures have been defined assuming that either the whole ensemble of realisations or, in the
ergodic case, the complete time history of one such realisation
is available. Under these conditions the classical relations between power spectrum, auto-correlation function, mean square value etc. are exact and devoid of any statistical interpretation
– perfect regularity has been found at least in the ideal situation. However, in practice the duration T of available record
remains finite and this is where the statistical aspect of data
analysis appears; the outcome $\hat{p}$ of any prescribed computational
algorithm aimed at determining some property p of the (whole)
random vibration will depend statistically on which interval
$(t, t + T)$ of the data has been chosen (more or less by chance).
To emphasize this statistical aspect, the chosen computation

yielding $\hat{p}$ is called an <u>estimator</u> and $\hat{p}$ itself an <u>estimate</u> of the population value p , whose determination would require the use of an infinite amount of data.

In a given situation involving random vibration the first step is to decide which property (or properties) p are physically relevant, e.g. the mean value $<x(t)>$, the r.m.s. value $<x^2(t)>^{1/2}$ the shape of the auto-correlation curve, the value of the spectral density $S(f)$ in a prescribed frequency range, or the shape of the probability density function $p(x)$ etc. The central problem of random data analysis then consists in setting up estimators with tolerable and more or less known statistical behaviour together with criteria for the sufficient (preferably minimum) amount of data in order to be able to decide when the difference between two estimates $\hat{p}_1$ and $\hat{p}_2$ based on two sets of data is greater than the expected statistical fluction if this data where drawn from the same random vibration, i.e. when the difference is statistically significant. Historically it is worth noting that the earliest estimator devised for the spectral density was set aside being useless for several decades because the periodogram as it was known proved to be statistically unsatisfactory. It has since been proved that in its original form the periodogram is an inconsistent estimator, i.e. its statistical scatter (variance) does not tend to zero as the data length T becomes infinite. About 1948–49 an alternative estimator was developed which involved the discovery of spectral smoothing as a way of decreasing the statistical scatter. With the increasing use of digital computers there has been a return to the smoothed periodogram together with a programme known as the NLOGN or FFT algorithm even though this method suffers from inflexibility in choice of bandwidth. An interesting account of this development is given by Tukey in ref. 17.1. The subroutine NLOGN can be found for instance in ref. 17.2. The point I am trying to make with the above historical account is that there is usually a host of possible estimators $\hat{p}_1, \hat{p}_2, \ldots$ of property p all based on a given set of data. Each of these estimators may be consistent. Some may be efficient in the sense that the expected statistical scatter is small, others may have advantage in ease of computation or implementation on an analogue device. Choosing a compromise is also the art of those analysing random data.

As an example of the difficulty involved in an a priori choice of an estimator consider the case where it is suspected that a set of measured amplitudes $x_1, x_2, \ldots, x_{2n+1}$ are random samples from a Gaussian amplitude distribution

$$p(x) \;=\; \frac{1}{\sqrt{2\pi}\,\sigma}\, \exp\left\{-(x-m)^2/2\sigma^2\right\}$$

with mean m and standard deviation σ . Both $\hat{m}_1 = x_{n+1}$ and $\hat{m}_2 = (x_1 + \ldots + x_{2n+1})/(2n+1)$ are consistent estimators of $m = E[x]$, but $\hat{m}_1$ has a standard deviation of 1.25 $\sigma/\sqrt{n}$ compared with only $\sigma/\sqrt{n}$ for $\hat{m}_2$ subject to the hypothesis of a Gaussian distribution function. A proof of the above is given in § 25.5 of ref.

17.3. In § 32.3, Ex. 1, of the same reference there is a proof that no other estimator of $\mathsf{E}[x]$ can have a smaller standard deviation than $\hat{m}_2$. This is what is meant by saying $\hat{m}_2$ is an _efficient estimator_ of the mean of a Gaussian distribution. However, if the amplitude x did not have the Gaussian probability density $\mathsf{p}(x)$ as assumed above but rather the Cauchy density

$$p(x) \;=\; \frac{1}{\pi\{1 + (x - m)^2\}} \;,$$

which also has a peak at $x = m$, the estimator $\hat{m}_2$ has the same scatter as x itself, whereas $\hat{m}_1$ has expected value m and scatter $(\pi/2)/\sqrt{n}$. In this case the efficient estimator would have scatter $\sqrt{2/n}$ i.e. 10% less than $\hat{m}_1$.

Since the behaviour of so many estimates depend on the form of $\mathsf{p}(x)$, it is logical to check the amplitude distribution early in the process of analysing random data.

In § 2.10 we saw that $<x>$, $<x^2>$ and $R(\tau)$ are defined in practice as time averages, and in § 3.5 it was explained that even the spectral density can be determined by the time average , $S_x(f) = B^{-1}<\tilde{x}^2>$, where $\tilde{x}$ is that part of the x at the output of a bandpass filter with cut-off points $f , f + B$. Even the estimation of $\mathsf{p}(x)$ usually involves a time average so it is logical to start with a consideration of time averages and their statistical properties.

17.2. Estimation of time properties

Figure 17.1 shows a simple mechanical time averaging device. Point A executes a random vertical motion $x(t)$, which obviously has either a peak in the power spectrum at zero frequency or a non-stationary mean $m(t)$ tending to a constant value as t in

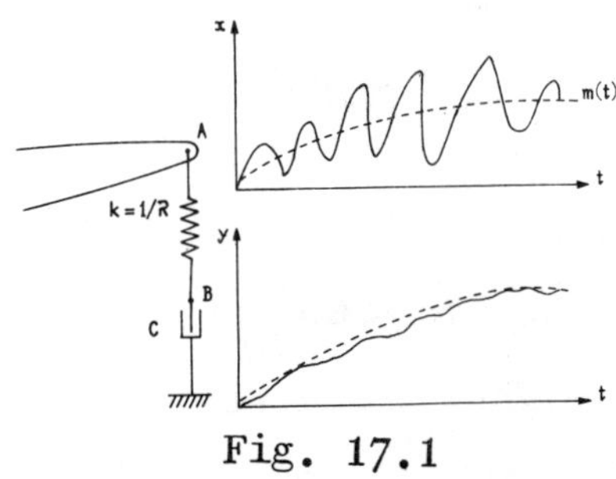

Fig. 17.1

creases. Let us denote the motion of point B of spring $k = 1/R$ damper c by $y(t)$. Clearly point B will not respond to the high-frequency motions of A but will follow the low-frequency trend in $x(t)$. The relation between x and y is

$$\dot{y} + \frac{1}{RC} y = \frac{1}{RC} x . \tag{1}$$

With $y(0)=0$, $x(t)=0$ for $t < 0$, and the notation 5.3 (4,6) we have

$$y(t) = \int_0^t \frac{1}{RC} x(s) W(t - s) ds = \frac{1}{RC} \int_0^t x(t - \tau) W(\tau) d\tau , \tag{2}$$

where $W(t) = \exp(-t/RC)$. It is usual to call $RC = t_0$ the filter time constant. The output $y(t)$ is called the exponentially-weighted past average of $x(t)$ and is written $<x(t)>_E$ i.e.

$$<x(t)>_E = \frac{1}{t_0} \int_0^t x(s) e^{-(t-s)/t_0} ds = \frac{1}{t_0} \int_0^t e^{-\frac{\tau}{t_0}} x(t - \tau) d\tau .$$

Taking the expectation of both sides of (3) yields

$$E[y] \;=\; \int_0^t t_0^{-1} E[x] \exp(-t/t_0) dt \;,$$

i.e. $\qquad\qquad E<x>_E \;=\; E[x]\{1 - \exp(-t/t_0)\}$

provided x is stationary. Thus, for $t > 5t_0$, $E<x>_E$ differs from $E[x]$ by less than 1%. The time average $<x>_T$ of x over $(t-T,t)$ would be obtained by using the weighting function

$$(4) \qquad\qquad W(t) \;=\; t_0/T \quad \text{for} \quad 0 < t < T.$$

Then (2) yields

$$(5) \qquad <x(t)>_T \;=\; \frac{1}{T}\int_{t-T}^{t} x(s) ds \;=\; \frac{1}{T}\int_0^T x(t-\tau) d\tau \;.$$

The implementation of (5) in order to compute $<x>_T$ in an analogue device requires an integrator to be reset after each time interval T. This has the disadvantage that the estimate of $<x>$ is only updated periodically, whereas the value of $<x>_E$ is available continuously. Although it is being assumed that $x(t)$ is stationary, a continuous estimate of the form $<x(t)>_E$ enables one to keep a check on this assumption.

There is a tendency today to accept the input $x(t)$ only at the discrete instants $t = 0, h, 2h, \ldots, nh, \ldots$ with a sample-hold device and denote the values $x(nh)$ by x_n. Expression (5) is then replaced by the approximation

$$\langle x_n \rangle_{T=rh} \;=\; \frac{1}{r}(x_n + x_{n-1} + \ldots + x_{n-r}) \tag{6}$$

called the running or summation average. This is an ordinary digital filter, because the number of weights is fixed. In the same way (3) is replaced by the approximation

$$\langle x_n \rangle_E \;=\; W_1 x_n + W_2 x_{n-1} + \ldots + W_n x_1 \tag{7}$$

which is called a recursive filter because the number of weights used increases with n and it is generated by the recurrence relation

$$\langle x_n \rangle_E \;=\; \frac{1}{N} x_n + \left(1 - \frac{1}{N}\right)\langle x_{n-1} \rangle_E \tag{8}$$

i.e. after each sampling interval h the value $\langle x_{n-1} \rangle_E$ in the store can be updated by multiplying it by the fixed factor $\left(1 - \frac{1}{N}\right)$ and adding $1/N$ times the new data value x_n to it. The meaning of N becomes evident if $\dot{y}$ in (1) is replaced by $(y_{n+1} - y_n)/k$, where $y_n = \langle x_n \rangle_E$. Hence

$$\langle x_{n+1} \rangle_E \;=\; \frac{h}{RC} x_n + \left(1 - \frac{h}{RC}\right)\langle x_n \rangle_E \tag{9}$$

i.e. $N = RC/h = t_0/h$ is the ratio of the filter time constant t_0 to the sampling interval h. This method of exponential averaging is stable without any physical upper limit on the time constant t_0 . In addition it is flexible since the value of N is easily changed. An improvement in the time to reach a steady estimate of $\langle x \rangle$ is obtained by starting with a small value of N which is gradually increased to the selected value. One manufacturer calls this a "quick look facility", and another describes the modified process as "true time averaging".

On the one hand, there is interest in keeping the averaging times T or $N = t_0/h$ as short as possible so that possible deviations from stationarity can be detected. On the other hand, these times must be long enough to reduce the statistical variations of the estimate $<x>_T$ or $<x>_E$ to a tolerably small amount. The variance (cf. § 2.3) serves as a convenient measure of the probable statistical error. For the time average $<x>_T$ we have

$$var<x>_T = \lim_{h \to 0} var<x_n>_{T=rh}$$

(10)

$$= \frac{2}{T} \int_0^T \left(1 - \frac{\tau}{T}\right) R_{\tilde{x}}(\tau) \, d\tau$$

where $\tilde{x} = x - \bar{x}$ so

$$R_{\tilde{x}}(\tau) = E\left[(x(0) - \bar{x})(x(\tau) - \bar{x})\right] = R_x(\tau) - \bar{x}^2$$

is the auto-correlation of $x(t)$ about its mean value $\bar{x}$.

Example

White noise in frequency band $0 < f < B$.

$$R_{\tilde{x}} = \sigma_x^2(\sin 2\pi B\tau)/2\pi B\tau \, , \, var<x>_T \doteq \frac{\sigma_x^2}{2BT} \, .$$

For the exponential average $<x>_E$ we have

$$var<x>_E = E\left[<x>_E^2\right] - \bar{x}^2 \, .$$

$$\overrightarrow{x} \quad \boxed{\alpha = \frac{1}{1 + i2\pi f t_0}} \longrightarrow y = <x>_E \ .$$

$$\text{averager}$$

The value of $E[y^2]$ is best calculated in the frequency domain:

$$E[y^2] \ = \ <y^2> \ = \ \int_0^\infty S_y(f)df \ = \ \int_0^\infty \frac{1}{1 + 4\pi^2 f^2 t_0^2} S_x(f)df \ .$$

Hence

$$\text{var} <x>_E \ = \ \int_0^\infty \frac{1}{1 + 4\pi^2 f^2 t_0^2} S_{\tilde{x}}(f)df \ . \tag{12}$$

From (10) and (12) we conclude that the values of $R_x(\tau)$ for small τ have a large effect on the fluctuations of $<x>_T$ whereas it is the low frequency values of $S_{\tilde{x}}(f)$ that affect the fluctuations in $<x>_E$.

Consider the concrete case

$$R_x(\tau) \ = \ a^2 e^{-k|\tau|} + \bar{x}^2 \ ,$$

$$S_{\tilde{x}}(f) \ = \ 4a^2 k/(4\pi^2 f^2 + k^2) \ .$$

Equations (10), (12) yield

$$\sigma_T^2 \ = \ \text{var} <x>_T \ = \ \frac{2a^2}{(kT)^2}(kT - 1 + e^{-kT}) < \frac{2a^2}{kT} \ ,$$

$$\sigma_E^2 \; = \; var <x>_E \; = \; \frac{a^2}{1 + kt_0} < \frac{a^2}{kT} \; .$$

Thus, for a specified r.m.s. error σ , we require

$$T > \frac{2a^2}{k\sigma^2} \quad \text{or} \quad t_0 > \frac{a^2}{k\sigma^2} \; .$$

If $k/2\pi = 100\,Hz$ and σ is to be less than 1% of the r.m.s. input a , we require

$$T > 32\,secs. \quad \text{or} \quad t_0 > 16\,secs.$$

i.e. for the given input exponential averaging is more efficient.

The concept of an averaging process as a low-pass filter is a useful one. Alternative filter receptances α have been proposed for reducing the statistical fluctuations in the average estimator for specified input spectra (see ref. 17.4).

17.3. Estimation of amplitude distribution

Subsequent data analysis can often be simplified if the amplitude distribution is known to be Gaussian. There are situations where the Gaussian nature can be inferred without resorting to measurements. For instance, if the random variable x is the sum of several <u>independent</u> random fluctuations, it is known that x will be very close to Gaussian even for a modest number of summands. The output of a linear system is always Gaussian when the input is Gaussian. On the other hand very little

is known about the effect of a linear system on a non-Gaussian
input; the simple relation $S_x = |\alpha|^2 S_p$ between input spec-
trum S_p and output spectrum S_x has no analogue for the amplitude probability distributions. The calculations in ref. 17.5
show that the outputs of two types of low-pass filter with time
constant t_0 become closer to Gaussian as the ratio t_0/T_m is increased when the input is one of a number of non-Gaussian types
including the random pulse signal described in § 6.6. However,
hopes for the tempting generalisation that the output of a low-
pass filter will always be approximately Gaussian are dashed by
the counter example of a (non-ergodic) non-Gaussian input which
produces non-Gaussian output irrespective of the value of t_0/T_m.
High-pass filters show no tendency at all to produce Gaussian
outputs.

If a process $\{x(t)\}$ is not ergodic, the only way
to calculate the amplitude distribution at a given time t is
to sample the amplitudes

$$^1x(t), \quad ^2x(t), \ldots, ^nx(t)$$

of n realisations of $\{x\}$ at time t. Let the superscripts be
chosen so that

$$^1x < {}^2x < {}^3x < {}^4x < \ldots < {}^nx .$$

The empirical distribution function $P_n(x)$ shown in Fig. 17.2
gives the number of observed amplitudes less than x . It is an

estimate of the true function $P(x)$. The statistical properties of

$$D_n \;=\; \max|P(x) - P_n(x)|$$

are well known and tabulated (e.g. see ref. 17.6). If we require $D_n < 0.04$ with probability 0.99 the tables yield $n > 1600$. For this value of n we can draw the corresponding confidence limits indicated by the dashed lines. If we were satisfied with a 95% confidence level, i.e.

$$Pr\{D_n < 0.04\} \;=\; 0.95$$

the necessary data would reduce to $n > 1160$.

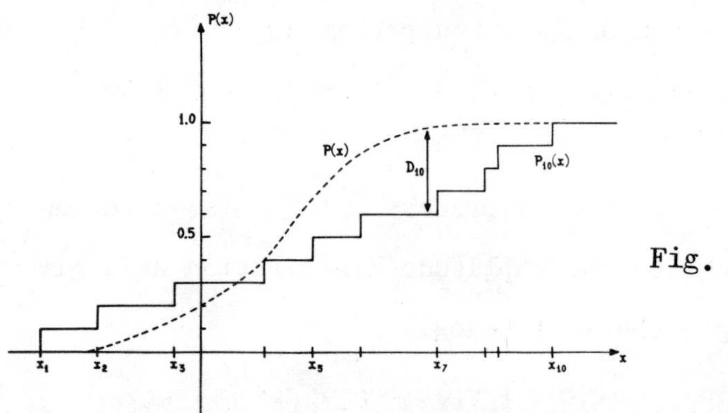

Fig. 17.2

If the process $\{x\}$ is ergodic and stationary, a single realisation $x(t)$ can be used to estimate the probability distribution function $P(x)$. This is much easier to do, as one simply has to form the ratio t_x/T, where t_x is the sum of the times for which $x(t) > x$ in the interval T (see Fig. 17.3). As a preparation to estimating the probable deviation of the statistic $P_T(x) = t_x/T$ from the true value $P(x)$ we note that the correlation time τ_c is defined such that $|R(\tau)| = 0.02\,R(0)$ when $\tau > \tau_c$.

If T were less than τ_c we would have very little information about the amplitude distribution, since the amplitudes are dependent on each other during this time. An expression for the variance of P_T can be found in ref. 17.7 together with the results of some approximate numerical computations. It is found that for $T = 15\,\tau_c$ the standard deviation of P_T could be as high as 10% of the standard devia

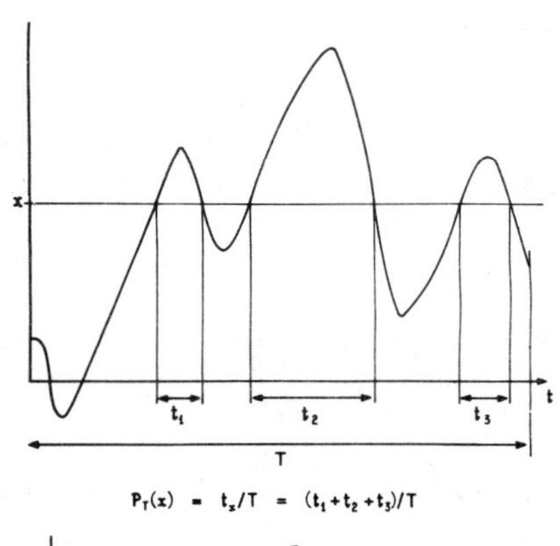

$$P_T(x) \;=\; t_x/T \;=\; (t_1 + t_2 + t_3)/T$$

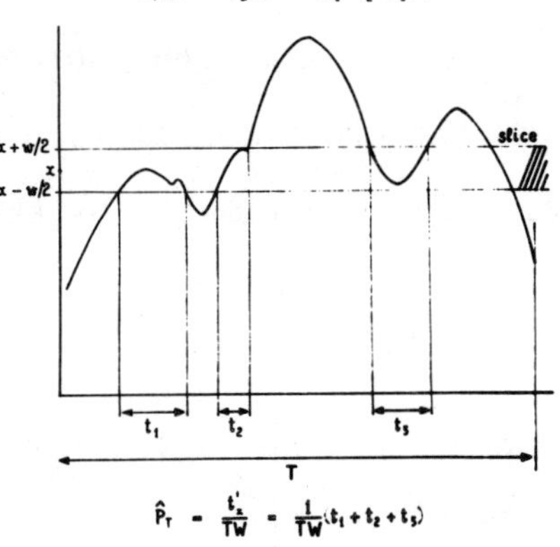

$$\hat{P}_T \;=\; \frac{t'_x}{TW} \;=\; \frac{1}{TW}(t_1 + t_2 + t_3)$$

Fig. 17.3

tion of x. The error drops to 4% for $T = 100\,\tau$, and even with $T = 700\,\tau_c$ nearly 2% error could be encountered.

The probability density $p(x) = P'(x)$ is estimated using a slicer which counts the time t'_x that x lies in the band $\left(x - \dfrac{W}{2},\, x + \dfrac{W}{2}\right)$ over the time interval T. Then we set

$$P_T \;=\; \frac{t'_x}{TW}\ .$$

A continuous estimate is found by forming a running average of P_T. A continuously displayed probability density (p.d.) is a good check on non-linearities in measuring equipment monitoring random vibrations, because any clipping of amplitudes shows up as a sharp peak. However, the probability density is poor as an indicator of extraneous sine wave (the p.d. of a sum is not the sum of the p.d.'s!). The autocorrelation or the power spectrum is used to detect extraneous signals.

17.4. Estimation of correlation functions

First consider the case where $\{x\}$ and $\{y\}$ are not ergodic. The cross-correlation $R_{xy}(t,t+\tau) = E[x(t)y(t+\tau)]$ defined in § 2.8 can be estimated by the ensemble average

$$(14) \qquad \hat{R}_{xy} = \frac{1}{n}\{{}'x(t)\,{}'y(t+\tau) + \ldots + {}^n x(t)\,{}^n y(t+\tau)\}$$

where ${}^i x(t)$ are the values at time t of n realisations of $\{x\}$ and ${}^i y(t+\tau)$ are the values at $t+\tau$ of n realisations of $\{y\}$. Since for each i we have

$$
\begin{aligned}
E[{}^i x(t)\,{}^i y(t+\tau)] &= \int_{-\infty}^{+\infty} {}^i x(t)\,{}^i y(t+\tau)\,p(x,t;y,t+\tau)\,dx\,dy \\
&= R_{xy}(t,t+\tau)
\end{aligned}
$$

$$(15)$$

it follows that $E[\hat{R}] = R$, i.e. $\hat{R}$ is an unbiased estimator of R . The statistical scatter of $\hat{R}$ about R for various values of n was first calculated by R.A. Fisher for the case when x and y are both Gaussian (see ref. 17.8). To get an idea of the size of sample n required, consider the case $R(\tau) = 0.8R(0)$.

With 95% confidence we expect

$$0.46\,R(0) < R < 0.93\,R(0) \quad \text{when} \quad n = 12$$
$$0.71\,R(0) < R < 0.86\,R(0) \quad \text{when} \quad n = 103.$$

For any sort of reasonable estimate $n = 100$ would seem a minimum. The stationarity of the process can be assessed by repeating the above calculations for different values of t .

Sampling across the ensemble is usually difficult (and time consuming), so there is great interest in using two time records $x(t)$, $y(t + \tau)$. An estimate of $R_{xy}(\tau)$ is then given by one of the time averages in § 17.2 of the product $x(t)y(t + \tau)$, e.g.

$$\hat{R}_{xy}(\tau) \;=\; <x(t)y(t + \tau)>_T \, . \tag{16}$$

However, the expression for the time variance analogous to 17.2 (10) is in general formidable. In fact we have

$$\text{var}\,\hat{R}_{xy} \;=\; \frac{1}{T^2}\int_0^T \left\{ E\!\left[x(u)y(u + \tau)x(v)y(v + \tau) \right] - R_{xy}^2(\tau) \right\} du\,dv \, . \tag{17}$$

For Gaussian white noise in $0 < f < B$ we have (see ref. 17.10, eq. (5.77))

$$\text{var}\,\hat{R}_{xy}(\tau) \;\doteqdot\; \frac{1}{2BT}\left\{ R_x(0)R_y(0) + R_{xy}^2(\tau) \right\} \, . \tag{18}$$

In particular,

$$\text{var}\,<x^2> \;=\; \sigma_x^4/BT \, . \tag{19}$$

From (18) it follows that var $\hat{R}$ decreases with lag τ . However, R usually decreases quicker, so the relative error usually increases considerably at large lag values.

An empirical approach to assessing the likely scatter in estimating R is to subdivide interval T into m subintervals and form m estimates of R . This would require $T > m\tau_c$ to be worthwhile.

The fact that the number of product-sum calculations in the estimator

$$(20) \qquad \hat{R}_{xy}(\tau) \;=\; \frac{1}{n} \sum_{i=1}^{n} x(t_i)y(t_i + \tau)$$

at n lag values increases as n^2 quickly makes it uneconomic on a universal computer. Robinson (ref. 17.2) reports that this is serious enough to justify equipping the SDS 9600 computer with a separate correlation and filter unit having direct access to the computer's memory, independent of the central processor.

An alternative method of speeding up the product-sum calculation is offered by the polarity coincidence correlator which calculates the estimator

$$(21) \qquad \tilde{r} \;=\; \frac{1}{n} \sum_{i=1}^{n} \tilde{x}(t_i)\tilde{y}(t_i + \tau) ,$$

where $\tilde{x}, \tilde{y} = \pm 1$ according as $\tilde{x}, \tilde{y} \gtrless 0$. Now

$$(22a) \qquad \lim_{n \to \infty} \tilde{r} \;=\; r \;=\; 1\Pr\{x > 0, y > 0\} + 1\Pr\{x < 0, y < 0\}$$

$$-1\Pr\{x > 0, y < 0\} - 1\Pr\{x < 0, y > 0\} \, . \qquad (22b)$$

When x , y are Gaussian with mean zero, we have

$$r \;=\; 4\int_0^\infty p(\xi,\eta)\,d\xi\,d\eta - 1 \qquad (23)$$

$$\;=\; \frac{4}{\pi}\arcsin(R_{xy}(\tau)/\sigma_x\sigma_y)\, . \qquad (24)$$

Thus $R_{xy}(\tau)$ is determined apart from the scale factor $\sigma_x\sigma_y$ by the value of r .(It is worth noting that as a consequence the frequency characteristic of a zero-mean Gaussian process is determined entirely by the times of its zero crossings). A successful commercial correlator developed in Britain exploits the idea of speeding up the multiply-add operations though amplitude quantization. Input x is quantized into the levels $0, \pm 1, \pm 2, \pm 4$ represented by a three-bit binary whereas y is finally quantized into seven binary bits. Mulitplication of pairs of x and y thus reduces to shifts of 0, 1 or 2 places in the y -register. This simplification is indeed necessary because up to $2^{17} = 131\ 072$ data points can be processed.

17.5. Estimation of power spectra

Expressions in Chapter 3 suggest the following two estimators of spectral density

$$\hat{S}_1(f) \;=\; \frac{2}{T}|A_T(if)|^2 \, ,$$

$$(26) \qquad \hat{S}_2(f) \; = \; \frac{1}{B} <\tilde{x}^2(t)> \; ,$$

where x denotes the output of a narrow-band filter with pass-band $(f, f+B)$ and input $x(t)$. Moreover $< \; >$ could be taken to be one of the time averages discussed in § 17.2. Noting that

$$\sigma_{\tilde{x}}^2 \; = \; \int_f^{f+B} S(f)df \; \approxeq \; BS(f) , \text{ it follows from } 18.4(19) \text{ that}$$

$$(27) \qquad \text{Var } \hat{S}_2 \; \approxeq \; \frac{1}{B^2} \text{Var} <x^2> \; = \; \frac{\sigma_x^4}{B^2 BT} \; = \; \frac{S^2(f)}{BT} \; .$$

Hence a decrease in bandwidth in order to get a less "smudged" picture of the spectrum leads to an increase in statistical error unless the amount of data T is also increased. The way in which the statistical error varies with frequency is rather irregular.

The large variance inherent in the so-called periodogram $\hat{S}_1$ proved an insurmontable obstacle for many years. Bartlett (ref. 17.9) showed that, if $A_T(if)$ is replaced by a finite Fourier series based on the n equally spaced samples $x_i = x(ih), i = 1,2,...,n,$ with $nh = T$, then

$$\hat{S}_1(f) \; = \; 2h \sum_{r=-n+1}^{r=n-1} \left(1 - \frac{|r|}{n}\right) \hat{R}_r \cos(2\pi rhf) \; ,$$

where

$$(28) \qquad \hat{R}_r \; = \; \frac{1}{n-r} \sum_{i=1}^{n-r} x_i x_{i+r} \quad (r=0,1,...,n-1) \; .$$

The weakness in this estimator is that $\hat{R}_r$ becomes very unreliable for large values of lag r . It can be shown that the equivalent bandwidth of this estimator is equal to T^{-1} . Thus, even as T is increased, the product BT in equation (27) remains equal to unity. Estimator $\hat{S}_1$ is inconsistent meaning that its variance does not tend to zero as $T \rightarrow \infty$. The idea suggested in ref. 17.9 for increasing the bandwidth and thereby reducing the statistical fluctuations in $\hat{S}_1$ was to subdivide the interval nh into say k equal subintervals each of length mh and to average the k corresponding auto-correlation functions. It was shown that this is equivalent to using the new estimator

$$\hat{S}_3(f) \;=\; 2h \sum_{r=-m+1}^{r=m-1} \left(1 - \frac{|r|}{m}\right)\hat{R}_r \cos(2\pi rhf) \tag{29}$$

which uses the estimated auto-correlation function over the restricted interval $mh = nh/k = T/k$. The equivalent bandwidth in the estimation is now k/T . The values of k chosen must represent a compromise between reducing statistical errors and increasing resolving power.

In digital data processing there is great interest in using an estimator based on $\hat{S}_1$, because the Fourier transform $A_T(if)$ can be calculated rapidly and directly from the data. The statistical misbehaviour of $\hat{S}_1$ is corrected to a certain extent by using "spectral smoothing". To see how this is already built into estimator $\hat{S}_3$, let $h \rightarrow 0$ set $k(\tau) = (1-k|\tau|/T) =$

$= 1 - |r|/m$ and take the expectation of both sides of (29)

$$E[\hat{S}_3] = 2\int_{-\infty}^{+\infty} k(\tau)R(\tau)e^{-i2\pi f\tau}\,d\tau\,,$$

$$= \int_{-\infty}^{+\infty} k(\tau)\int_{-\infty}^{+\infty} S(g)e^{i2\pi g\tau}\,dg\,e^{-i2\pi f\tau}\,d\tau\,,$$

Since $R(\tau) = \dfrac{1}{2}\int_{-\infty}^{+\infty} S(g)e^{i2\pi g\tau}\,dg\,.$

If the Fourier transform of $k(\tau)$ is denoted $K(f)$ we have

$$K(f) = \int_{-\infty}^{+\infty} k(\tau)e^{-i2\pi f\tau}\,d\tau$$

(Note that k is called the lag window and K the spectral windows). Now

$$E[\hat{S}_3] = \int_{-\infty}^{+\infty} S(g)\int_{-\infty}^{+\infty} k(\tau)e^{-i2\pi(f-g)\tau}\,d\tau\,dg$$

$$E[\hat{S}_3(f)] = \int_{-\infty}^{+\infty} S(g)K(f-g)\,dg\,.$$

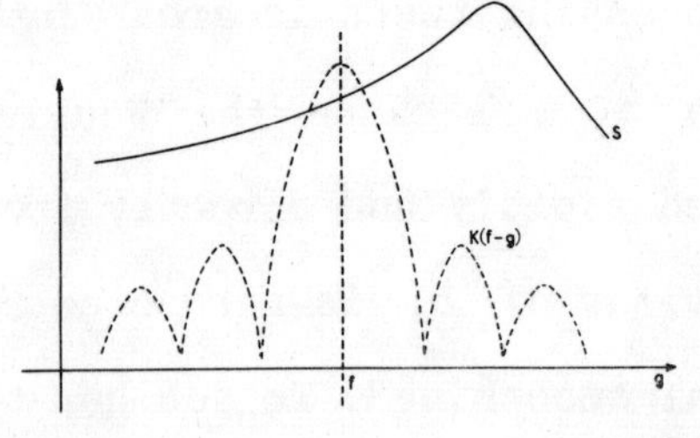

This convolution integral is to be interpreted as if the expected value of estimate $\hat{S}_3$ is obtained from the true spectrum S by looking at it trough the window $K(f-g)$.

When estimating S at point f , a neighbouring peak corresponding to one of the side-lobes of K can seriously distort the estimate. This is called "leakage". As an instance, the slightest non-zero mean in the data gives rise to a delta function in S at $f = 0$. Through leakage this can affect estimates of S over quite a range of low frequencies. The latest developments have been directed at finding modified estimators whose spectral windows exhibit smaller side-lobes. One of these suggests multiplying the first and last 10% of data points by a cosine "roll-off factor" thus introducting the so-called data window.

One commercial spectrum analyser comes to grips with the statistical variations by storing up to 256 estimates based on separate intervals of record and averaging these. This additional device is called an ensemble averager.

An important practical consideration in choosing a method of spectral analysis is whether the bandwidth should be constant or variable over the full frequency range. In the field of acoustics a system of internationally standardised third-octave filters is used, i.e. the band-width is approximately proportional to centre frequency. Spectra computed on a proportional bandwidth look better on a log-grequency plot whereas with a fixed bandwidth a linear scale seems more natural.

Bibliography for Chapter 17

17.1   Harris B.(Editor): <u>Spectral Analysis of Time Series</u>.
        John Wiley & Sons Inc., 1967.

17.2   Robinson E.A.: <u>Multichannel Time Series Analysis with
        Digital Computer Programs</u>.
        Holden-Day, San Francisco, 1967.

17.3   Cramér H.: <u>Mathematical Methods of Statistics</u>.
        Princeton University Press, Princeton 1961.

17.4   Middleton D.: <u>An Introduction to Statistical Communica-
        tion Theory</u>.
        McGraw-Hill, New York, 1960.

17.5   McFadden J.A.: Probability Density of Output of a Filter
        when the Input is a Random Telegraphic Signal.
        IRE trans. Information Theory, 1959.

17.6   Birnbaum Z.W.: Numerical Tabulation of the Distribution
        of Kolmogorov's Statistic.
        J. Amer. Stat. Assocn., Vol. 47, p.425, 1952.

17.7   Baburin V.M.: Calculation of the Distribution Function
        for Random Processes from Experimental Data.
        Automation and Remote Control, Vol. 23, p.519,
        1962.

17.8   Fisher R.A.: <u>Statistical Methods for Research Workers</u>.
        Oliver and Boyd, Edinburgh, 1954.

17.9   Bartlett M.S.: Periodogram Analysis of Continuous Spec-
        tra.
        Biometrica, Vol. 37, pp. 1-16, 1950.

17.10   Bendat J.S. and Piersol A.G.: Analysis and Measurement
        Procedures.
        John Wiley and Sons, New York, 1971.

BIBLIOGRAPHY

1957
Eringen, A.C.: Response of Beams and Plates to Random Loads.
J. Appl. Mech. 24, pp. 46–52.

Thomson, W.T. & Barton, M.V.: The Response of Mechanical Systems
to Random Excitations. J. Appl. Mehc. 24, pp. 2
248–51.

1958
Bendat, J.S.: Principles and Applications of Random Noise Theory. Wiley, New York.

Coleman, T.L., Murrow, H.N. & Press, H.: Some Structural Response . Characteristics of a Large Flexible
Swept-Wing Airplane in Rough Air. J. Aero. Sci.
25 pp. 515–21.

Crandall, S.H. (Ed.): Random Vibration. MIT Press/Wiley, H.York.

Davenport, W.B. & Root, W.L. Random Signals and Noise. McGraw-
Hill, New York.

Mains, R.M.: Minimising Damage from Random Vibration. JASA 30
pp. 1127–29.

Morrow, C.T.: Averaging Time and Data Reduction Time for Random
Vibration Spectra. JASA 30 pp. 456–61 & 572–78.

Powell, A.: On the Approximation to the Infinite Solution by
the Method of Normal Modes for Random Vibrations.
JASA 30 pp. 1136–39.

Powell, A.: On the Fatigue Failure of Structures Due to Vibrations Excited by Random Pressure Fields. JASA
30 pp. 1130–35.

Samuels, J.C. & Eringen, A.C.: Response of a Simply Supported
        Timoshenko Beam to a Purely Random Gaussian
        Process. J. Appl. Mech. 25, pp. 496–500.

1959
Blackman, R.B., and Tukey, J.W.: The Measurement of Power Spec
        tra. Dover.

Caughey, T.K.: Response of a Nonlinear String to Random Loading:
        J. Appl. Mech. 26 pp. 341–344.

Caughey, T.K. Response of Van Der Pol's Oscillator to Random
        Excitation: J. Appl. Mech. 26 pp. 345–48.

Dyer, I.: Response of Plates to a Decaying and Convecting Ran-
        dom Pressure Field: JASA 31 pp. 922–28.

Etkin, B.: A Theory of the Response of Airplanes to Random
        Atmospheric Turbulence. J. Aerospace Sci. 26
        pp. 409–20.

Samuels, J.C. & Eringen, A.C.: On Stochastic Linear Systems.
        J. Math. & Phys. 38 pp. 85–103.

1960
Ariaratnam, S.T.: Random Vibrations of Nonlinear Suspensions:
        J.M.E.S. 2 pp. 195–201.

Bogdanoff, J.L. & Goldbert, J.E.: On the Bernoulli Euler Beam
        Theory with Random Excitation: J. Aerospace
        Sci. 27 pp. 371–76.

Caughey, T.K.: Random Excitation of a System with Bilinear
        Hysteresis: J. Appl. Mech. 27 pp. 649–52.

Caughey, T.K. Random Excitation of a Loaded Nonlinear String:
        J. Appl. Mech. 27 575–78.

Kats, I.I. & Krasovskii, N.N.: On the Stability of Systems
        with Random Parameters: P.M.M.(J. Math. Mech.)
        24 pp. 809–23.

Lyon, R.H.: Equivalent Linearisation of the Hard Spring Oscil-
          lator: JASA 32 pp. 1161-62.

Lyon, R.H.: On the Vibration Statistics of a Randomly Excited
          Hard Spring Oscillator: JASA 32 pp. 716-19.

Morrow, C.T.: Random Vibration: JASA 32 pp. 742-48.

Samuels, J.C.: On the Stability of Random Systems and the Stab-
          ilisation of Deterministic Systems with Random
          Noise: JASA 32 pp. 594

1961
Beer, F.P.: On the Response of Linear Systems to Time-Dependent
          Multidimensional Loading. J.Appl. Mech. 28 pp.
          50-55.

Caughey, T.K. & Dienes, J.K.: Analysis of a Nonlinear First-
          Order System with a White Noise Input: J. Appl.
          Phys. 32 pp. 2476-79.

Caughey, T.K. & Stumpf, H.J.: Transient Response of a Dynamic
          System Under Random Excitation: J. Appl. Mech.
          28 pp. 563-66.

Curtis, A.J. & Boykin, T.R.: Response of Two Degree of Freedom
          Systems to White Noise Excitation: JASA 33 pp.
          653-63.

Katz, I.I. & Krasovaskii, N.N.: On the Stability of Systems
          with Random Parameters: P.M.M. 25 pp. 809-23.

Kaufman, S. Lapinski, W.L. & McCaa, R.C.: Response of a Single
          Degree of Freedom Isolator to a Random Distur-
          bance: JASA 33 pp. 1108-12.

Kozin, F.: On the Probability Densities of the Output of Some
          Random Systems: J. Appl. Mech. 28 pp. 161-64.

Lyon, R.H. Heckl, M. & Hazelgrove, C.B.: Narrow Band Excitation
          of the Hard Spring Oscillator: JASA 33 pp. 1404-
          11.

Maidanik, G.: Use of Delta Function for the Correlations of Pres̲
           sure Fields: JASA $\underline{33}$ pp. 1598–1605.

Samuels, J.C.: Theory of Stochastic Linear Systems with Randomly
           Varying Parameters: JASA $\underline{33}$ pp. 1782–86.

Slepian, D.: First Passage Problem for a Particular Gaussian
           Process. Ann. Math. Stat. $\underline{32}$ pp. 610–12.

1962

Ariaratnam, S.T.: Response of a Loaded Nonlinear String to Ran-
           dom Excitation. J. Appl. Mech. $\underline{29}$ pp. 483–85.

Bogdanoff, J.L. & Kozin, F.: Moments of the Output of Linear
           Random Systems. JASA $\underline{34}$ pp. 1176–

Caughey, T.K. & Dienes, J.K.: The Behaviour of Linear Systems
           with Random Parametric Excitation: J. Math. &
           Phys. $\underline{41}$ pp. 300–10.

Chelpanov, I.B.: Vibration of a Second Order System with a Ran-
           domly Varying Parameter: P.M.M. $\underline{26}$ pp. 1145–52.

Clarkson, B.L.: The Design of Structures to Resist Jet Noise
           Fatigue. J. Roy. Aero. Soc. $\underline{66}$ pp.603–613.

Clarkson, B.L. & Ford, R.D.: The Response of a Typical Aircraft
           Structure to Jet Noise: J. Aero. Soc. $\underline{66}$ pp. 31–
           40.

Crandall, S.H.: Random Vibration of a Nonlinear System with a
           Set Up Spring: J. Appl. Mech. $\underline{29}$ pp. 477–82.

Crandall, S.H. & Yildiz, A.: Random Vibration of Beams: J. Appl.
           Mech. $\underline{29}$ pp. 267–75.

Hasselmann, K. Random Excitation of Vibrating Systems: ZAMM $\underline{42}$
           pp. 465–76.

Heckl, M.A.: Vibrations of Point Driven Cylindrical Shells: JASA
           $\underline{34}$ pp. 1553

Lin, Y.K.: Stresses in Continuous Skin Stiffener Panels Under
          Random Loading: J Aerospace Sci. 29 pp. 67

Longuet-Higgins, M.S.: The Distribution of the Intervals between
          Zeros of a Stationary Random Function: Phil. Trans.
          Roy. Soc. A. 254 pp. 557-99.

Lyon, R.H. & Maidanik, G.: Power Flow Between Linearly Coupled
          Oscillators: JASA 34 pp. 623

Maidanik, G.: Response of Ribbed Panels to Reverberent Acoustic
          Fields: JASA 34 pp. 809

Rosenblüth, E. & Bustamente, J.: Distribution of Structural Res-
          ponse to Earthquakes. J. Eng. Mech. Div., ASCE
          88 EM3 pp. 75-106.

Slepian, D.: The One-Sided Barrier Problem for Gaussian Noise:
          B.S.T.J. 41 pp. 463-501.

Smith, P.W.: Response and Radiation of Structural Modes Excited
          by Sound: JASA 34 640

Smith, P.W.: Response of Nonlinear Structures to Random Excita-
          tion: JASA 34 pp. 827

Spence, H.R. & Luhrs, H.N.: Structural Fatigue Under Combined
          Random and Swept Sinusoidal Vibration: JASA 34
          pp 1076

Tack, D.H. & Lambert R.F.: Response of Bars and Plates to Bound-
          ary Layer Turbulence: J. Aerospace Sci. 29 pp.
          311-22.

Thomson, W.T.: Continuous Structures Excited by Correlated Ran-
          dom Forces: Int. J. Mech. Sci. 4 pp. 109-14.

Weidenhammer, F.: Vibration of Foundations Under Random Excita-
          tion: Ing. Arch. 31 pp. 433-43.

Winter, E.F. & Bies, D.A.: Correlation Properties of Flexural
          Waves in Long Thin Bars: JASA 34 pp. 472-75.

1963

Caughey, T.K.: Derivation and Application of the Focker Planck
          Equation to Discrete Nonlinear Dynamic Systems:
          JASA 35 pp. 1683-

Caughey, T.K.: Equivalent Linearisation Techniques. JASA 35 pp.
          1706

Crandall, S.H.: Perturbation Techniques for Random Vibration of
          Nonlinear Systems: JASA 35 pp. 1700

Crandall, S.H.: Zero Crossings, Peaks, and Other Statistical
          Measures of Random Responses: JASA 35 pp. 1693

Crandall, S.H.(Ed): Random Vibration Vol. 2 MIT Press, N. York.

Crandall, S.H. & Mark, W.D.: Random Vibration in Mechanical Sys
          tems. Academic Press. New York.

Jullien, Y.G.: The Exact Solution of the Response of a Damped
          Rod to Random Excitation: J. Appl. Mech. 30 pp.
          312

Kozin, F.: On Almost Sure Stability of Linear Systems with Ran-
          dom Coefficients: J. Math. & Phys. 43 pp. 59-67.

Liebowitz, M.A.: Statistical Behaviour of Linear Systems with
          Randomly Varying Parameters: J. Math. Phys. 4
          pp. 852-58.

Lin, Y.K.: Application of Non-Stationary Shot Noise in the Study
          of System Response to a Class of Non-Stationary
          Excitations: J. App. Mech. 30 pp. 555-59.

Lin, Y.K.: Probability Distributions of Stress Peaks in Linear
          and Nonlinear Structures: AIAA J. 1 pp. 1133

Lin, Y.K.: Nonstationary Response of Continuous Structures to
          Random Loading: JASA 35 pp. 222

Lyon, R.H.: Empirical Evidence for Nonlinearity and Directions
          for Future Work: JASA 35 pp. 1712

Robson, J.D.: An Introduction to Random Vibration. Edinburgh
            University Press.

Samuels, J.C.: The Dynamics of Impulsively and Randomly Varying
            Systems: J. Appl. Mech. $\underline{30}$ pp. 25–30.

Shimogo, T.: Nonlinear Vibration of Structures Under Random
            Loading: Bull. J.S.M.E. $\underline{6}$ pp. 44–52.

Shimogo, T.: Unsymmetrical Nonlinear Vibration Systems Under
            Random Loading: Bull. J.S.M.E. $\underline{6}$ pp. 53–59.

Smith, P.W. & Maime, C.I.: Fatigue Tests of a Resonant Struc-
            ture with Random Excitation: JASA $\underline{35}$ pp. 43

Trubert, M.R.P.: Response of Elastic Structures to Statistical-
            ly Correlated Multiple Random Excitations: JASA
            $\underline{35}$ pp.

1964
Barret, J.F.: The Use of Characteristic Functionals and Cumulant
            Generating Functionals to Discuss the Effect of
            Noise in Linear Systems: J. Sound Vib. $\underline{1}$ pp. 229–
            38.

Bozich, D.J.: Spatial Correlation in Acoustic–Structural Coupl-
            ing: JASA $\underline{36}$ pp. 52

Chiesa, A.: Experimental Studies on Noise Inside Cars: J. Sound
            Vib. $\underline{1}$ pp. 211–25.

Crandall, S.H. Khabbaz, G.R., & Manning, J.E.: Random Vibration
            of an Oscillator with Nonlinear Damping: JASA
            $\underline{36}$ pp. 1330

Goldberg, J.E., Bogdanoff, J.L. & Sharpe, D.R.: The Response of
            Simple Nonlinear Systems to a Random Disturbance
            of the Earthquake Type: Bull. Seism. Soc. Amer.
            $\underline{45}$ pp. 263

Herbert, R.E.: Random Vibrations of a Nonlinear Elastic Beam:
        JASA 36 pp. 2090

Khabbaz, G.R.: Significance of the Cross-Correlation Between
        the Modes of a Structure on its Response: AIAA
        J. 2 pp. 2211

Klein, G.H.: Random Excitation of a Nonlinear System with Tan-
        gent Elasticity Characteristics: JASA 36 pp. 2095

Lin, Y.K.: Random Vibration of a Myklestad Beam: AIAA J. 2 pp.
        1448

Lin, Y.K.: On Nonstationary Shot Noise: JASA 36 pp. 82-84.

Lyon, R.H. & Eichler, E.: Random Vibration of Connected Struc-
        tures: JASA 36 pp. 1344

Lyon, R.H. & Maidanik, G.: Statistical Methods in Vibration
        Analysis: AIAA J.2 pp. 258-69.

Weidenhammer, F.: Stabilitätsbedingungen fürSchwinger mit zu-
        fälligen Parameterregungen: Ing. Arch. 33 pp.
        404-15.

White, P.H.: Effect of Boundary Flexibility on the Response of
        a String to Convected Random Loading: JASA 36
        pp.

1965
Bogdanoff, J.L. & Citron, S.J.: Experiments with an Inverted
        Pendulum Subject to Random Parametric Excitation:
        JASA 38 pp. 447

Caughey, T.K. & Gray, A.H., Jr.: On the Almost Sure Stability
        of Linear Dynamic Systems with Stochastic Coef-
        ficients: J. Appl. Mech. 32 pp. 365-72.

Clarkson, B.L. & Mercer, C.A: Use of Cross Correlation in Study-
             ing the Response of Lightly Damped Structures to
             Random Forces: AIAA J. 3 pp. 287-91.

Eichler, E.: Thermal Circuit Approach to Vibrations in Coupled
             Systems and the Noise Reduction of a Rectangular
             Box: JASA 38 pp. 995

Gray, A.H.: Behaviour of Linear Systems with Random Parametric
             Excitation: JASA 37 pp. 235

Herbert, R.E.: Random Vibrations of Plates with Large Amplitudes:
             J. Appl. Mech. 32 pp. 547-52.

Lin, Y.K.: Nonstationary excitation and Response in Linear Sys-
             tems Treated as Sequences of Random Pulses: JASA
             pp. 453

Maestrello, L.: Measurement of Panel Response to Turbulent Bound
             ary Layer Excitation: AIAA J. 3 pp. 359-61.

Mercer, C.A.: Response of a Multi-Supported Beam to a Random
             Pressure Field: J. Sound Vib. 2 pp. 296-306.

Newland, D.E.: Energy Sharing in the Random Vibration of Nonlin-
             earity Coupled Modes: J. Inst. Math. Appl. 1 pp.
             199.

Pal'mov, V.A.: Thin Shells Acted on by Broad Band Random Loads:
             J. Appl. Math. Mech. 29, pp. 905-13.

Pretlove, A.J.: Bond Stresses in a Randomly Vibrating Sandwich
             Plate: Multi-Modal Theory. J. Sound. Vib. 2 pp.
             1-22.

Rice, J.R.: Starting Transients in the Response of Linear Sys-
             tems to Stationary Random Excitation: J. Appl.
             Mech. 32 pp. 200-01.

Roberts, J.B.: On the Harmonic Analysis of Evolutionary Random
             Vibration: J. Sound Vib. 2 pp. 336-52.

Roberts, J.B.: The Response of Linear Vibratory Systems to Random Impulses: J. Sound Vib. $\underline{2}$ pp. 375-90.

Roberts, J.B. & Bishop, R.E.D.: A simple Illustration of Spectral Density Analysis: J. Sound Vib. $\underline{2}$ pp. 37-42.

Robson, J.D. & Roberts, J.W.: A theoretical Basis for the Practical Simulation of Random Motions: J.M.E.S. $\underline{7}$ pp. 264-51.

1966

Beer, F.P. & Ravera R.J.: Effect of Spacewise Variations in a Random Load Field on the Response of a Linear System:AIAA J. $\underline{4}$ pp. 1651-54.

Bendat, J.S. & Piersol, A.G.: Measurement and Analysis of Random Data: Wiley, New York.

Broch, J.T.Some Aspects of Sweep Random Vibration: J. Sound Vib. $\underline{3}$ pp. 195-204.

Chisholm, C.J.: Random Vibration Techniques Applied to Motor Vehicle Structures: J. Sound Vib. $\underline{4}$ pp. 129-36.

Coupry, G.: Measurement of the Spectral Densities of Turbulence by a Method Deduced from Rice's Formulae: J. Sound Vib. $\underline{4}$ pp. 123-28.

Crandall, S.H. Chandiramani, K.L. & Cook, R.G.: Some First-Passage Problems in Random Vibration: J. Appl. Mech. $\underline{33}$ pp. 532-38.

Felszeghy, S.F.,& Thomson, W.T.: Probability Distribution of Bilinear System Response to Impulse Excitation: J. Appl. Mech. $\underline{33}$ pp. 384-87.

Gräffe, P.W.U.: Stability of a Linear Second Order System under Random Parametric Excitation: Ing. Arch. $\underline{35}$ pp. pp. 202-05.

Gray, A.H., JR.: First Passage Time in a Random Vibrational System: J. Appl. Mech. 33 pp. 187-91.

Jullien, Y.: Random Vibrations of Elastic Spherical Shells: J. de Mécanique 5 pp. 419-38.

Karnopp, D.: Coupled Vibratory System Analysis Using the Dual Formulation: JASA 40 pp. 380-84.

Karnopp, D. & Scharton, T.: Plastic Deformation in Random Vibration: JASA 39 pp. 1154-61.

Nemat-Nasser, S.: On the Response of Continuous Media to Random Excitations: Int. J. Solids & Struc. 2 pp. 371-84.

Newland, D.E.: Calculation of Power Flow Between Coupled Oscillators: J. Sound Vib. 3 pp. 262-76.

Piersol, A.G.: The Development of Vibration Test Specifications for Flight Vehicle Components: J. Sound Vib. 4 pp. 38-115.

Pullen, C.L. & Peterson, H.C.: Spectral Analysis of the Transient Response of a System to Random Excitation: J. Appl. Mech. 33 pp. 700-02.

Ribner, H.S.: Response of a Flexible Panel to Turbulent Flow: Running Wave versus Modal Density Analysis: JASA 40 pp. 721-26.

Rice, J.R. & Beer, F.P.: First Occurrence Time of High Level Crossings in a Continuous Random Process: JASA 39 pp. 523-35.

Roberts, J.B.: The Response of a Simple Oscillator to Band-Limited White Noise: J. Sound Vib. 3 115-26.

Roberts, J.B.: On the Response of a Simple Oscillator to Random impulses: J. Sound Vib. 4 pp. 51-62.

Robson, J.D.: The Random Vibration Response of aSystem having Many Degrees of Freedom. Aero Quart. 17 pp. 21-30.

Wang,P.K.C.: On the Almost Sure Stability of Linear Stochastic
          Distributed Parameter Dynamical Systems: J. Appl.
          Mech. 33 pp. 182-86.

Weidenhammer, F.: Stabilitätsbedingungen für Schwingungsysteme
          mit zufälliger Parametererregung durch weisses
          Rauschen: Ing. Arch. 35 pp. 1-9.

1967
Caughey, T.K. & Dickerson, J.R.: Stability of Linear Dynamic
          Systems with Narrow Band Parametric Excitation:
          J. Appl. Mech. 34 pp. 709-13.

Chen, T.C.: Response of a Linear System to Non Stationary Shot
          Noise: JASA 41 pp. 822-26.

Dimentberg, M.F.: Determining the Statistical Characteristics
          of a Linear Dynamic System from Measurements of
          its Motion: J. Appl. Math. Mech. 31 pp. 1092-97.

Dimentberg, M.F.: Subharmonic Resonance in a System with a Ran-
          domly Varying Natural Frequency: J. Appl. Math.
          Mech. 31 pp. 761-62.

Gray, A.H. Jr.: Frequency-Dependent Almost Sure Stability Condi-
          tions for a Parametrically Excited Random Vibra-
          tional System: J. Appl. Mech. 34 pp. 1017-19.

Gray, A.H., Jr.: Some Energy Bounds for Randomly Excited Beams:
          JASA 41 pp. 615-17.

Haines, C.W.: Hierarchy Methods for Random Vibrations of Elastic
          Strings and Beams: J. Eng. Maths. 1 pp. 293-306.

Janssen, R.A., & Lambert, R.F.: Numerical Calculation of Some
          Response Statistics for a Linear Oscillator under
          Impulsive Noise Excitation: JASA 41 pp. 827-35.

Karnopp, D.: Power Balance Method for Nonlinear Random Vibra-
          tion: J. App. Mech. 34 pp. 212-14.

Karnopp, D. & Brown, R.N.: Random Vibration of Multi Degree of
         Freedom Hysteretic Structures: JASA 43 pp. 54-59.

Lin, Y.K.: Probabilistic Theory of Structural Dynamics: Mc Graw
         Hill, New York.

Lyon, R.H.: Random Noise and Vibration in Space Vehicles. Shock
         & Vibration Information Center U.S. Department
         of Defense.

Lyon, R.H.: Spatial Response Concentrations in Extended Struc-
         tures. J. Eng. Ind. (ASME Series B) 89 pp. 754-8.

Maestrello, L. Use of Turbulent Model to Calculate the Vibration
         and Radiation Responses of a Panel: Sound Vib. 5
         pp. 407-48.

Milacic, V.R. & Gartner, J.R.: The Application of Correlation
         Theory for the Investigation of Cutting Torque
         in Horizontal Milling: Int. J. Mech. Tool Des.
         7 pp. 391-407.

Nevel'son: Behaviour of a Linear System under Small Random Exci-
         tation of its Parameter: J. Appl. Math. Mech.
         31 pp. 552-55.

Novak, M.: A statistical Solution of the Lateral Vibration of
         Cylindrical Structures in Air Flow: Acta Technica
         CSAV 12 pp. 375-404.

Patrick, T.J.: Sweep Sine Wave Simulation of Random Vibration
         and its Effect on Design with Particular Reference
         to Space Rockets: J. Sound Vib. 5 pp. 37

Priestley, M.B.: Power Spectral Analysis of Non Stationary Ran-
         dom Processes: J. Sound Vib. 6 pp. 86-97.

Shapiro, A.: An Analysis of the Pseudo- Autocovariance Function
         as a Technique for Signal Echo Detection: J. Sound
         Vib. 5 pp. 93-99.

Shinozuka, M. Sato, Y.: On the Numerical Simulation of Nonstationary Random Processes: Proc. ASCE (J. Eng. Mech. Div.) 93 EM 1 11-40.

Shinozuka, M. Yao, J.T.P.: On the Two-Sided Time-Dependent Barrier Problem: J. Sound Vib. $\underline{6}$ pp. 98- 104.

Soovere, J. & Clarkson, B.L.: Frequency Response Functions from Cross Correlation: Bartlett Weighting Function. AIAA J. $\underline{5}$ pp. 601-03.

Srinivasan, S.K. Subramanian, R. & Kumaraswamy, S.: Response of Linear Vibratory Systems to Non-Stationary Stochastic Impulses: J. Sound Vib. $\underline{6}$ pp. 169-79.

Trubert, M.: Structural and Electromechanical Interaction in the Multiple Excited Technique for Random Vibration Testing: JASA $\underline{41}$ pp. 1185-92.

Tung, C.C.: The Effects of Runway Roughness on the Dynamic Response of Airplanes: J. Sound Vib. $\underline{5}$ pp. 164-72.

Ungar, E.E.: Statistical Energy Analysis of Vibrating Systems: J. Eng. Ind. (ASME Sec. B.) $\underline{89}$ pp. 626-32.

Usher, T., Jr.: Average Control for Sinusoidal and Random Vibration Testing: JASA $\underline{41}$ pp. 840-49.

Wendeborn, J.O.: Description of Runway Roughness by Power Spectra: Autobiltechnische Zeitschrift $\underline{69}$ pp. 117-18.

Wolf, A.A.: An Ergodic Theorem and its Generalisation: J. Franklin Inst. pp. 283-99.

1968
Ariaratnam, S.T. & Sankar, T.S.: Dynamic Snap Through of Shallow Arches under Stochastic Loads: AIAA J. $\underline{6}$ pp. 798-802.

Atkinson, J.D. & Caughey, T.K.: Spectral Density of Piecewise Linear First Order Systems Excited by White Noise:

Int. J. Non-lin. Mech. $\underline{3}$ pp. 137-56.

Barnoski, R.L.: The Maximum Response to Random Excitation of
Distributed Structures with Rectangular Geometry:
J. Sound Vib. $\underline{7}$ pp. 333-50.

Barnoski, R.L.:On the Single Highest Peak Response of a Dual
Oscillator to Random Excitation: J. App. Mech.
$\underline{35}$ pp. 414-16.

Chapman, C.P.: Scanning Techniques for Random Noise Testing: J.
Environmental Sci. $\underline{11}$ pp. 27-35.

Clarkson, B.L.: Stresses in Skin Panels Subjected to Random A-
coustic Loading: Aero. Journ. $\underline{72}$ pp. 1000-10.

Foster, E.T.Jr.: Semilinear Random Vibration in Discrete Systems:
J. Appl. Mech. $\underline{35}$ pp. 560-64.

Fryba, L.: The Inverse Problem in Stochastic Processes: JAMM $\underline{48}$

Hammond, J.K.: On the Response of Single and Multi-Degree of
of Freedom Systems to Stationary Random Excitations: J. Sound Vib. $\underline{7}$ pp. 393-416.

Hedin, G.L. & Lambert, R.F.: Numerical Prediction of Response
Peak Statistics of Linear Systems Excited by Impulsive Noise. JASA $\underline{43}$ pp. 1319-23.

Houbolt, J.C.: Exceedances of Structural Interaction Boundaries
for Random Excitation: AIAA Journ. $\underline{6}$ pp. 2175-82.

Infante, E.F.: On the Stability of Some Linear Non-Autonomous
Systems: J. Appl. Mech. $\underline{35}$ pp. 7-12.

Iwan, W.D.& Lutes, L.D.: Response of the Bilinear Hysteretic
System to Stationary Random Excitation: JASA $\underline{43}$
pp. 545-52.

Jacobson, M.J.: Stress and Deflection of Honeycomb Panels Loaded by Spatially Uniform White Noise: AIAA Journ.
$\underline{6}$ pp. 1503-10.

Knowles, J.K.: On the Dynamic Response of a Beam to Random Excitation: J. Appl. Mech. $\underline{35}$ pp. 1-6.

Lepore, J.A. & Shah, H.C.: Dynamic Stability of Axially Loaded Columns Subjected to Stochastic Excitation: AIAA Journ. $\underline{6}$ pp. 1515-1521.

Liepins, A.A.: Digital Computer Simulation of Railroad Freight Car Packing: Trans ASME (B) (J. Eng. Ind.) $\underline{90}$ pp. 707-7.

Mercer, C.A. & Hammond, J.K.: On the Representation of Continuous Random Pressure Fields at a Finite Set of Points: J. Sound Vib. $\underline{7}$ pp. 49-61.

Nemat-Nasser, S.: On the Response of Shallow Thin Shells to Random Excitation: AIAA Journ. $\underline{6}$ pp. 1327-30.

Piersol, A.G.: Study of the Interaction of Combined Static and Dynamic Loads. J. Sound Vib. Vib. $\underline{7}$ pp. 319-32.

Roberts, J.B.: An approach to the First Passage Problem in Random Vibration: J. Sound Vib. $\underline{8}$ pp. 301-28.

Robson, J.D.: Deductions from the Spectra of Vehicle Response Due to Road Profile Excitation: J. Sound Vib. $\underline{7}$ pp. 156-58.

Scharton, T.D. & Lyon R.H.: Power Flow and Energy Sharing in Random Vibration: JASA $\underline{43}$ pp. 1332-41.

Smits, T.I. & Lambert, R.F.: System Reliability Prediction in Impulsive Noise Environments Based on Wear Dependent Failure Rates Using Response Peak Statistics: JASA $\underline{43}$ pp. 1344-50.

Stanisic, M.M.: Response of Plates to Random Load: JASA $\underline{43}$ pp. 1351-57.

Thomas, J.H.: Random Vibrations of Thin Elastic Plates: Z.A.M.P. $\underline{19}$ pp. 921-26.

Yamada, Y. & Takemiya, H.: Studies on the Responses of Multi-
        Degrees of Freedom Systems Subjected to Random
        Excitation with Applications to the Tower and
        Pier Systems of Long Span Suspension Bridges:
        Mem. Fac. Eng. Kyoto Univ. 30 pp. 371-96.

1969
Barnoski, R.L. & Maurer, J.L.: Mean-square Response of Simple
        Mechanical Systems with Parametric Excitation:
        J. Appl. Mech. 36 pp. 221-27.

Clinch, J.M.: Measurement of the W 11 Pressure Field at the Sur
        face of a Smooth Walled Pipe Containing Turbulent
        Water Flow: J. Sound Vib. 9 pp. 398-419.

Crocker, M.J.: The Response of a Supersonic Transport Fuselage
        to Boundary Layer and to Reverberant Noise: J.
        Sound Vib. 9 pp. 6-20.

Crocker, J.M. & Price A.J.: Sound Transmission using Statistic-
        al Energy Analysis: J. Sound Vib. 9 pp. 469-86.

Crocker, J.J. & White, R.W.: Response of Uniform Beams to Homo-
        geneous Random Pressure Fields: JASA 45 pp. 1097-
        1103.

Field, C.F.: Cross-Correlation and Milling Machine Dynamics -
        A Case Study: Int. J. Mech. Tool Des. 9 pp. 81-
        96.

Gersch, W.: Average Power and Power Exchange in Oscillators: J.
        Acon. Soc. Am. 46 pp. 1180-85.

Hwang, C. & Pi W.S.: Random Acoustic Response of a Cylindrical
        Shell: AIAA J. 7 pp. 2204-10.

Liu, S.C. & Jhaveri, D.P.: Spectral Simulation and Earthquake
        Site Properties: Proc. ASCE (J. Eng. Mech. Div.)
        95 EM5 pp.1145-68.

Lyon, R.H.: Statistical Analysis of Power Inspection and Res-
        ponse in Structures and Rooms: J. Acon. Soc.
        Am. 45 pp. 545-65.

Roberts, J.B.: Estimation of the Probability of First-Passage
        Failure for a Linear Oscillator: J. Sound Vib.
        pp. 42-61.

Stearn, S.M.: Measurement of Correlation Coefficients of Accel-
        eration of a Randomly Excited Structure: J. Sound
        Vib. 9 pp. 21-27.

Strawderman, W.A. & Brand, R.S.: Turbulent Flow-Excited Vibration
        of a Simply-Supported Rectangular Flat Plate:
        JASA 45 pp. 177-92.

White, D.J.: Effect of Truncation of Peaks in Fatigue Testing
        using Narrow-Band Loading: Int. J. Mech. Sci.
        11 pp. 667-75.

White, D.J. & Lewszuk, J.: Narrow Band Fatigue Testing with
        Amsler Vibrophore Machines: J. Mech. Eng. Sci.
        11 pp. 598-604.

White, P.H.: Cross-Correlation in Structural Systems: JASA 45
        pp. 1118-28.

Yang, J.N. & Shinozuka, M.: Numerical Fourier Transform in Ran-
        dom Vibration: Proc. ASCE (J. Eng. Mech. Div.)
        95 EM3 pp. 731-46.

Zeman, J.L. & Bogdanoff, J.L.: A Comment on Complex Structural
        Response to Random Vibrations: AIAA J. 7 pp. 1225-
        31.

1970
Clinch, J.M.: Prediction and Measurement of the Vibrations In-
        duced in Thin-Walled Pipes by the Passage of
        Internal Turbulent Water Flow: J. Sound Vib. 12
        pp. 429-51.

Crandall, S.H.: Distribution of Maxima in the Response of an
        Oscillator to Random Excitation: JASA 47 pp.838-
        -845.

Hart, G.C.: Building Dynamics due to Stochastic Wind Forces:
        Proc. Am. Soc. Civ.Eng. (J. Struc. Div.) Vol. 96
        No. ST3, pp. 535-50.

Man, F.T.: On the Almost Sure Stability of Linear Stochastic
        Systems: J.Appl. Mech. 37 pp. 541-43.

Mark, W.D.: Spectral Analysis of the Convolution and Filtering
        of Non-Stationary Stochastic Processes: J. Sound
        Vib. 11 pp. 19-64.

Ping-I-Wang, A.: A note on the Random Responses of a Thin Cylin-
        drical Shell: J. Sound Vib. 12 pp. 393-96.

Srinivasan, S.K. & Kumaraswamy, S.: Characteristic Functional
        of Nonstationary Shot Noise: J. Appl. Mech. 37
        pp. 543-44.

Williams, T.R.G.: Fatigue Under Sinusoidal and Narrow-Band Ran-
        dom Conditions: J. Sound Vib. 11 pp. 251-61.

CONTENTS

Page